NATO and the Mediterranean

NATO and the Mediterranean

Edited by
Lawrence S. Kaplan, Robert W. Clawson,
and Raimondo Luraghi

With Essays by

Victor Alba
Thomas H. Etzold
Mariano Gabriele
Lawrence S. Kaplan and Robert W. Clawson
Bruce R. Kuniholm
Alberto Li Gobbi
Raimondo Luraghi
André Martel and Claude Carlier
Albano Nogueira
Joaquin Oltra
S. Victor Papacosma
Stevan K. Pavlowitch
Antonio Pelliccia
William H. Rowden
E. Timothy Smith

SR Scholarly Resources Inc.
Wilmington, Delaware

First published 1985
Printed and bound in the United States of America

Scholarly Resources Inc.
104 Greenhill Avenue
Wilmington, Delaware 19805

Library of Congress Cataloging in Publication Data
Main entry under title:

NATO and the Mediterranean.

Includes index.
1. North Atlantic Treaty Organization—Mediterranean Region—Addresses, essays, lectures. 2. Mediterranean Region—Defenses—Addresses, essays, lectures.
I. Kaplan, Lawrence S. II. Clawson, Robert W.
III. Luraghi, Raimondo. IV. Title: N.A.T.O. and the Mediterranean.
UA646.5.M4N38 1984 355′.031′091821 84-16055
ISBN 0-8420-2221-X

Contents

Preface and Acknowledgments

The Lyman L. Lemnitzer Center has published two volumes in the NATO series: *NATO After Thirty Years* (1981) and *The Warsaw Pact: Political Purpose & Military Means* (1982). The present book is a product of collaboration with the Centro Studi sulla Difesa of the University of Genoa. As such, it involved the solving of problems familiar to any transnational effort, including financial, linguistic, and postal obstacles. Fortunately, the special relationship between the two centers has helped to surmount most of these successfully. The result is a volume reflecting not only the scholarship of specialists but also in some instances the personal evaluations of military and political officials of members of the alliance. All told, nationals from six NATO countries were represented as speakers, and representatives of Greece and Turkey were present as commentators.

This book, like the conference itself, has many debts to acknowledge. Among the more prominent are those organizations that have made financial and logistical contributions to the project. These include the Consiglio Nazionale delle Ricerche, the Italian Ministry of Defence for supplying transportation, the U.S. Information Agency, the NATO Information Directorate, and the Rockefeller Foundation. The hosts of the conference—the staff of the Rockefeller Foundation's Bellagio Study and Conference Center at the Villa Serbelloni in Bellagio, notably Roberto Celli, Angela Barmettler, and Marta Hill—were especially supportive. Our thanks also to Susan E. Garfield of the Rockefeller Foundation's New York Office.

The authors have all benefited from the comments and criticism offered by the following participants who sat with them at the long rectangular table in the Villa Serbelloni: William Barnes, Robert Bemis, Antonello Biagini, Anton Giulio De Robertis, Joseph Favorite, Robert Fisher, M. R. Frisby, Paul F. Goldberg, Elias Katsareas, Flora Lewis, Ali Engin Oba, Alberto Santoni, and Gilbert M. Sauvage.

Special mention must be made of Mark R. Rubin whose services included program planner, translator, transportation director as well as on-site manager and commentator.

The staffs of the two centers eased many of our editorial burdens. In Italy they included Margherita Mazza. In the United States, Ruth Young managed much of the enterprise as well as performed vital secretarial services, Wilda Lunceford once again gave the center the benefit of her typing skills, and Nancy Brendlinger provided essential Italian language assistance as did James Miller of the Department of State Historical Office.

To all those who shared in this project, the editors give their thanks.

Lawrence S. Kaplan
Robert W. Clawson
Raimondo Luraghi

Introduction

EAST-WEST COMPETITION IN THE MEDITERRANEAN

To ask why the southern flank of NATO—the Mediterranean basin—has been a relatively neglected area of study is to draw attention to the question whether the area possesses sufficient cohesiveness to justify a common strategy. In other words, is there a Mediterranean, as opposed to a southern, flank of NATO?

Historical evidence can offer a negative impression on two geographical fronts. On the east, there are severe regional problems in the enduring conflicts between Greece and Turkey which divert the attention of the two parties from the alliance as well as from a Mediterranean theater. On the west, Spain and France appear so far removed from the east that they seem to have difficulty in finding a commonality of concerns. Even the formal identification of the Mediterranean was made only after considerable difficulties had been overcome. Italy in 1949, and Greece and Turkey three years later, were given grudging acceptance by their European partners. As LAWRENCE KAPLAN and ROBERT CLAWSON observe in their overview of the problem, the command of this disparate and discordant flank was itself the subject of bitter dispute over authority on the part of American, British, and French military leaders. The result, reflected in the array of overlapping commands, was confusion and apparent impotence.

NATO, however, had too much at stake to allow dissolution of the southern flank. The survival of the alliance could hang on the steadfastness of the Mediterranean, particularly if the Soviets were able to play successfully on Greek-Turkish rivalry, or upon the anti-Western sentiments of the Arab littoral nations of the Mediterranean, or upon the return of post-Tito Yugoslavia to the Warsaw Pact fold. Moreover, the Mediterranean as an arena of the Cold War is more volatile than the central or northern flanks, not only because of internal disturbances within the region but also because the mobility of the Soviet navy poses special dangers not found elsewhere. Its blue water ships move from the Black to the Mediterranean seas with few hindrances and frequently with disturbing support from disaffected Mediterranean nations, some of them in NATO itself.

While the major interests of both NATO and the Warsaw Pact organization may center in Germany, the periphery can spark a conflict that could engulf both organizations. It is this particular threat that Kaplan and Clawson see as the major reason for the importance of the U.S. Sixth Fleet, the primary agent for stability in the eastern Mediterranean.

ADMIRAL WILLIAM ROWDEN underscores this point by listing crises from the 1940s to the present, a time during which the Sixth Fleet has used its superiority on the sea to affect events on the shores. Had they ended in failure for the West, Europe would have been severely damaged by the loss of vital goods and resources supplied them through Mediterranean shipping. The stakes are almost as high for the Soviet Union, given the fact that the majority of its imports and exports pass through the Bosphorus. Unlike the West, however, Soviet naval power, in defense of its interests and in pursuit of advantage, has expanded in the past generation. Even if the Soviets also face problems, particularly those stemming from lack of control of choke points, the burden of response has fallen to the Sixth Fleet.

Rowden observes that in one vital area no alternative exists, namely, in the West's relations with the southern and eastern littorals of the Mediterranean. Although Libya and Lebanon are outside the boundaries of NATO and hence outside its defense responsibilities, actions in those areas have direct effect on the security of the allies. Without a NATO mandate, only the Sixth Fleet is in a position to serve the organization as an arm of American naval power, but, he suggests, there are limits to its capabilities; the burdens must be shared more equally by the NATO allies in and out of the Mediterranean sector. It is in the interest of central and northern members to give the southern flank naval and air reinforcements needed to maintain stability for the entire organization.

THOMAS ETZOLD shares some of these judgments, although from a different observation. He too finds the "out-of-area" problems a key to the volatility of the area, from which there is no escape. To have the United States respond to them unilaterally, however, risks a reorientation of American priorities from northern and Central Europe to the Middle East as well as a diversion of American materiel from the central to the southern flank. For the health of the alliance in all its aspects, the Mediterranean deserves the attention and support of all its members.

While he also notes the rising Soviet naval power over the past decade, Etzold sees its role as defensive, if not quite marginal. Its successes have been limited, and the importance of the Mediterranean is well beneath that of Central Europe in its scale of priorities. Given the perennial turmoil in the southern flank, the Soviets will take advantage of trouble where they can, but its offensive posture is only tactical. Strategically the position is defensive, with its primary concerns centered on the putative dangers of U.S.-Israeli collaboration and on its own excessive fears of Western encirclement. If the role is primarily political, the major challenge to NATO lies not in Soviet power but in the disordered relations among the allies, particularly in their relations with Southwest Asia.

THE NATO SOUTHERN FLANK

GENERAL ANTONIO PELLICCIA focuses on another aspect of the allies' internal disarray: the fallacious conceptualization of the role of air power in the defense of the southern sector. Much of the problem derives from a land-bound organization of military forces, whereby Europe is divided into sectors, or flanks, countering a potential enemy on the east in each flank. This traditional approach to defense is underscored, he believes, by the domination of army figures in the SHAPE command; Lauris Norstad was the only SACEUR to be selected from the air force.

Unfortunately, according to General Pelliccia, the proper role of the air force cannot be confined to rigid military categories; it serves more than an ancillary function of helping armies to destroy military installations or industries. Primarily it should concentrate on controlling airspace. The Soviet Union, which has secured most of the traditional tsarist objectives, sees this use of air power as an opportunity to complete its goal of access to the outer world through warm water ports. The powerful array of sophisticated Soviet aircraft bears witness to the means it believes to be vital in a nuclear age. Belatedly, NATO is coming to a recognition of this question, as it gives greater meaning to interoperability of air weaponry among allies previously tied to mutually incompatible systems.

GENERAL ALBERTO LI GOBBI shares with General Pelliccia an assumption that the internal disarray among the allies in the Mediterranean is as much a problem for NATO as the Soviet threat from the outside. Ostensibly, the most difficult problems lie in the eastern Mediterranean where Turkey borders a hostile Soviet Union on the north and an unstable Arab Moslem world on the south. Additionally, Turkey contends with Greece over long-standing mutual grievances in Cyprus and the Aegean Sea. The major weakness, however, may be found in Italy to the west, presumably further removed from the Warsaw bloc and traditionally a major bulwark of NATO in the Mediterranean. Li Gobbi warns that persistent Italian unwillingness to share defense burdens, commensurate with its size and with the stake in the common security, could tempt an enemy toward adventures that might not be undertaken otherwise. He couples this weakness with the potential presence of Soviet troops on the northeast border of Italy should Yugoslavia return to the Soviet bloc in the future.

MARIANO GABRIELE also centers much of his attention on the role of Italy in the Mediterranean, but his emphasis is less on the military activities, or on the naval presence, than on Italy's part, along with other Mediterranean powers, in entering into economic commitments to the North African and Middle Eastern nations. Long-run stability requires a relationship that would serve the alliance as well as the Third World powers involved.

Professor Gabriele does not minimize the Soviet threat, particularly its naval elements, but he does point out that there are limits to military influence, specifically naval, in light of changes in technology. This applies to the Soviet as well as to NATO forces. His attention, however, is to the out-of-area peripheries of the Mediterranean, reaching down to Sudan on the south and Iran on the east. Indeed, he suggests that these areas point to the recasting of SHAPE's geographic scope, with the Mediterranean becoming a new central area and the more stable German front serving as a northern sector of NATO. On the military side, NATO allies should complement and accept some of the responsibilities that have fallen primarily on the United States, as they have increasingly in the Middle East. Together all the partners should become more sensitive to the underlying economic and social issues that are as much a source of destabilization as military weakness.

THE IBERIAN CONNECTION

Portugal's role as a Mediterranean power requires the same kind of broad conceptualization that was used in 1949 to consider Italy an Atlantic power. Its primary justification would lie in its character as an Iberian nation sharing with Spain the far western shores of the Mediterranean. But even this is complicated by the obvious fact that until 1982 Spain was not a member of NATO and indeed was a pariah under Franco to most of NATO's constituency. AMBASSADOR ALBANO NOGUEIRA emphasizes the consistent longstanding Portuguese connection with England which so often in the past helped to determine Portuguese actions in 1949. Indeed, he notes the semantic agility used by the dictator Salazar in his conduct in the 1930s and 1940s to reconcile his obvious ideological support for fascism, and presumably nazism, with a loyalty to the British alliance. No matter how specious it may appear, Salazar used the British connection to maintain a neutrality that served his own interests during the war.

In entering NATO, as Ambassador Nogueira observes, Portugal abandoned its neutrality but not its ideological stance or the British alliance. Salazar's former sympathy for fascism was transformed into a deepening anticommunism, and his links to England were reinforced by the expanding role of the United States in European affairs. If Salazar needed to justify his deviation from an old isolationism from Europe, Europeans needed to look the other way when talking of democracy and freedom in the company of the dictator, but the mutual interests of all parties dictated Portugal's membership. In the absence of Spain, NATO's entry to the Mediterranean, as well as its access to the South Atlantic, had been secured by the charter membership of one of the two Iberian powers in the alliance.

Although Spain has a more obvious part to play than Portugal on the southern flank of NATO, VICTOR ALBA points out the special features distinguishing that Iberian country from both Portugal and the rest of Europe. Paramount among them has been the long tradition of isolationism that has made the Pyrenees a barrier to Spain's integration with the West. What is noteworthy then about Spain's distance from NATO was the ardent interest Generalissimo Franco had in identifying his nation with the Atlantic Alliance. This had been the distinguishing element in Spanish external relations. Spain's rejection by the allies was the product of NATO revulsion against Franco's Fascist regime.

The historical antecedents of Spain's membership in NATO devolves on the special relationship with the United States, particularly after the Korean War seemed to make Spanish bases vital to the security of the rest of Europe. Much of the ambivalence post-Franco Spain has toward NATO derives from the anti-Americanism nourished by Spanish anti-Fascists over the thirty years of a de facto Hispano-American alliance. The United States symbolized the alliance and, as a major prop of the Franco regime, became an ongoing target of the Left. Alba emphasizes, however, countervailing forces, such as the legitimization of the new republic by its association with mostly democratic nations of the organization, that prevailed in order to bring Spain into the alliance. As for the allies, the inclusion of Spain would solidify the defense of the southwestern Mediterranean but also could blunt the sensitive Anglo-Spanish controversy over the future sovereignty of Gibraltar. Alba's own hope is that NATO would serve Spain by weaning it away from excessive dependence on the United States and facilitating its links to the European Community.

While JOAQUIN OLTRA is skeptical about the very existence of Iberia, his carefully drawn picture of the immediate future of Spain and NATO and his interpretation of an interlocking past for Spain and Portugal belie his professed skepticism. Over the centuries there has been a symbiotic Portuguese-Spanish relationship, expressed as much in collaboration as in conflict. Most notably, Portugal's acceptability to the NATO allies served as a conduit for Spain's relationship with the United States, and the bonds of neutrality of the two dictatorships in World War II spared the peninsula the horrors of that war.

Neutralism, or more aggressively, isolationism, has a long history in Spain. If that nation departed from it after the death of Franco, it was largely because of the need for European respectability that membership in NATO would confer upon the rightist parties in the post-Franco era. Whether this new departure would survive the arrival to power of a left-leaning Socialist government is all the more a question since opposition to NATO became an article of faith for much of the Left in the recent past. A promised referendum on the subject could undo the ties made in 1982 and make impossible further integration into the organization. While recognizing the irony of the Socialists' position on Gibraltar, formerly the preserve of the Right, Oltra also points to the sobering responsibility of power. The González government realizes the need for NATO, in company with Socialist governments of other NATO nations. It is unlikely that Spain will move toward military integration in the immediate future, but it is even less likely that it would abandon the alliance itself.

FRANCE, ITALY, AND NORTH AFRICA

For ANDRÉ MARTEL and CLAUDE CARLIER the Mediterranean is both destiny and opportunity. Unlike any NATO member, with the exception of Spain, France has both an Atlantic and a Mediterranean coast. While the latter should have made an important difference, the historic concentration on the Atlantic and the Rhine obscured it. Among the many contributions of de Gaulle was the Fifth Republic's rediscovery of the Mediterranean, where France presumably could play a special role.

The attractiveness of the Mediterranean was all the greater because there France could compensate for the Anglo-Saxon domination of the alliance and the consequent northern orientation of NATO. When France withdrew its naval forces from NATO's jurisdictional control in the Mediterranean, it did not signal an abandonment of the area. Rather, it permitted France to mount its own defense of the western Mediterranean, complementary of NATO's and perhaps all the more effective because of its separation from SHAPE. Martel and Carlier emphasize as well the opportunity that the Mediterranean geopolitical posture of France created for a special relationship with North Africa and the Middle East. They believe that the economic and strategic importance of the Mediterranean today invites France to play a predominant role that would serve the interests of the West by limiting Soviet influences in the out-of-area shores of the Mediterranean.

Given the central role of Italy in the defense of the Mediterranean over the past generation, it is noteworthy that Italy, like the Mediterranean itself, was given only marginal consideration in the formulation of the North Atlantic Treaty in 1948–49. As E. TIMOTHY SMITH points out, there was little initial support for its membership among the framers. Italy was outside the Atlantic boundaries, restricted by treaty provisions in its armaments, and torn by internal dissension over its relationship with the Atlantic powers.

If the allies were won over to an Italian role, it was primarily the work of American planners, and to a lesser extent of French negotiators concerned with including their Algerian departments in the organization. Algeria focused its attention on the Mediterranean, and Italy benefited from this attention. The American interest was less in the potential strategic importance of the Italian peninsula, or even in its position in the defense of Central Europe; these would come to the fore, as Smith notes, only after NATO evolved into a military organization. Instead, Americans pressed for Italian inclusion to minimize dangers of Communist subversion, particularly from the Communist Party of Italy which had been poised for an electoral victory in 1948.

In his commentary, RAIMONDO LURAGHI restates an ironic conundrum that turns up in many of the articles; namely, the coexistence of NATO's neglect of the southern flank with its flickering recognition that the survival of other sectors of Europe depends on the security of the Mediterranean. The security of the southern sector can be traced from west to east, according to Luraghi, beginning with Turkey as the first line of defense, Greece as second, and Italy as third. These bulwarks suffer from serious shortcomings, independent of the intra-NATO struggle of Greeks and Turks. The dangers in one sense are quantifiable: outmoded weapons in Turkey; insufficient funds in Italy. In another, they devolve from structural weaknesses outside NATO's control.

Much of the successful Soviet presence in the Mediterranean derives from its use of out-of-area countries, such as Syria and Libya, that provide port facilities to the Soviet navy. Fittingly for a scholar observing the scene from an Italian perspective, the prospect of a Yugoslav union with the Warsaw bloc inspires a nightmare of Soviet troops facing Nuova Gorizia. This would be a disaster for all NATO, not simply for the defense position of Italy. Yet Luraghi's tone is not resolutely pessimistic. He sees not only the failures and weaknesses of the Soviet forces but also the potential gains in a closer unity among the allies. While there is no single solution to a series of problems, one approach certainly is to sensitize the NATO community to the vital importance of the southern sector.

THE BALKAN POWERS

With all the attention understandably placed on Greece and Turkey by the alliance, the relationship of a nonmember, Yugoslavia, has always loomed large on the NATO horizon, as STEVAN PAVLOWITCH observes. Thirty years ago it was a potential model for schism in the Communist world and a fellow traveler at least of NATO through membership in the Balkan Pact. Since Tito's death it has been the subject of more concern in the defense of the Mediterranean than the Aegean or any other out-of-area problem. The reason is the specter of a Yugoslav reconciliation with the Soviets, resulting in Soviet troops on the Italian border. Dangerous as the divisions between Greece and Turkey or the Israeli-Arab conflict have been to the stability of the Mediterranean, they do not invoke images of Soviet soldiers in the area as does post-Tito Yugoslavia.

This concern was the salvation of Tito's regime, as he played on Western fears and Soviet hopes. While Pavlowitch credits Tito's balancing act for the successes he achieved, the author identifies vital faults in Tito's postures. For one, his ideological weakness for Soviet communism made his heirs and him susceptible to a mistaken judgment that American imperialism was more dangerous than the Soviets. While he knew better pragmatically, his weapon to keep the balance—nonalignment—was faulty and is weaker today than ever before. In post-Tito Yugoslavia, economic chaos, as well as ethnic divisiveness and ideological kinship, would be the elements to bring the country back to the Soviet fold, not Soviet armies. Not that Pavlowitch believes this is likely to be Yugoslavia's future; the role of the military alone might thwart it.

What is disturbing has been NATO's hope of continuing Titoism without Tito, as if this were the only hope for Yugoslavia's independence, and as if Titoism had been the best solution for Yugoslavia's relationship with the West. Genuine nonalignment requires a support for change, a more pluralistic system that need not be anti-Soviet in character. Given Yugoslavia's wish to see NATO continue its containment of Soviet power, one way to resolve Yugoslavia's isolation might be a new Balkan connection that would undergird its independence from both blocs.

If the Mediterranean is difficult to conceive as a unit, the peculiar history of the Balkan region helps to account for it. Its troubled past, Pavlowitch observes, has left a legacy that has kept the area as much of a tinderbox in the last decades of the twentieth century as it had been in the early part of the century. NATO has been a unifying element, or at least could have served as the instrument for coordinating the latent strength of the region harmoniously within a larger European framework. For a time during the Korean War when SHAPE was created, such a role seemed possible. Indeed, for a brief period in the 1950s the Balkan Pact, linking Yugoslavia with Greece and Turkey, appeared to fulfill its potential. This fragile unity shattered over Yugoslavia's confidence in dealing with a post-Stalin Soviet Union as well in the outbreak, primarily over Cyprus, of conflict between the NATO allies Greece and Turkey.

For Greece since the 1950s, NATO and its primary member, the United States, have replaced Britain and France as unreliable protectors. Once more Greece was dependent upon a great power and not simply against Soviet dominance but against its major enemy, Turkey. Again, the benefactor appeared to betray the client. The judgment of Greece has been that, in the struggle between Greece and Turkey, the United States has valued and continues to place higher importance on its Turkish connection.

AS S. VICTOR PAPACOSMA observes, Greece appears to struggle with its NATO partners more vigorously than with the Soviet antagonist. On the one hand, there is the continuing conflict with Turkey that centers on the future of Cyprus but is by no means confined to this issue. Disputes continue over operational control, territorial waters, and airspace rights in the Aegean. On the other hand, resentment against American official tilting toward the Turkish side outweighs satisfaction of Greek support in American domestic politics. The future of American bases in Greece is not secure.

Nonetheless, there is a strain of continuity in Papacosma's account of Greece's position in NATO. The response of Greece to its frustrations was not permanent estrangement from the United States or NATO. As he explains, under Papandreou, as well as under his predecessors, there is no place else for Greece to turn. While the temptation to play off the Warsaw Pact against NATO is present today, and indeed has been acted upon, there are limits to Greece's independence. One major factor is the advantage accruing to Turkey within NATO if Greece leaves the alliance. At best a position similar to Romania's in the Warsaw Pact would be open to Greece. As long as Greece regards Turkey as its enemy and continues to fear its incursions in every area of the Aegean, NATO and the southeastern sector are inevitably weakened.

While the Turkish links with NATO have been subjected to periods of tension, they have been less tempestuous than the NATO-Greek passions. An important part of the explanation, according to BRUCE KUNIHOLM, has been the abiding Turkish fear of the Soviet Union as a hostile neighbor. Not even gestures of appeasement, if not friendship, after Stalin's death or in the late Brezhnev years have altered Turkey's basic position. In this sense its position is not that of a mirror image of Greece, which was and is a rival but not the primary foe. Kuniholm maintains that there is widespread recognition of mutual dependence on Turkey's part that makes cooperation in the eastern Mediterranean of vital importance for the Balkan nations and all their NATO partners.

Turkish loyalty to the alliance has been accompanied by long-standing American support. This has been weakened by congressional sympathy for the Greek side in the Cyprus quarrel and by congressional attachment to what may be an unrealistic 7-to-10 ratio in U.S. aid to Greece and Turkey. Similarly, Turkey has been an outcast in recent years in the European Community because of its authoritarian regime. Kuniholm, however, sees little prospect of neutralism or accommodation to Soviet pressures; the Russian presence is too oppressive. At the same time, the strategic position of Turkey has given it a position of priority in the eastern Mediterranean, both in American and NATO planning. The bonds that anchored Turkey to NATO over thirty years ago are still tight.

APPENDIX

In the comments presented by the chiefs of staff of the Italian military services, there are a number of themes that touch on and underscore the contributions of the foregoing essays. The chiefs' message is directed to the political authorities of their own country as well as to their allies in NATO. To the former, they call attention to inadequate defense budgets; to the latter, they ask for "due attention," as General Umberto Cappuzzo suggests, to the importance of Italy as the major link between the central and southern flanks of the alliance.

Each of the chiefs of staff also raises questions regarding the role of the Mediterranean in NATO strategy. As do many of the authors, they observe that the sea is not merely a southern front, similar in nature but inferior in significance, to the central front. Instead, it is an area of special dangers stemming from the Soviet presence as well as special opportunities for NATO to address a relationship with the Third World that may have important consequences for the future of the alliance.

Contributors

VICTOR ALBA is professor emeritus of political science at Kent State University and is the author of over thirty books on Spanish and Latin American politics. His most recent publications are *Transition in Spain* (1980) and *The Communist Party in Spain* (1983). He now lives in Spain.

CLAUDE CARLIER earned his doctorate from the University of Montpellier, France. He is associate director of the Institute of Contemporary Military History in Paris and teaches defense studies and aerospace history at the Institute of Political Studies at Aix-en-Provence, the University of Montpellier, the Sorbonne, the Air College, and the Air War College. He is the author of two important works in his field: *Le Développement de l'aéronautique militaire française de 1958 à 1970* (1979) and *L'Aéronautique française 1945–1975* (1983).

ROBERT W. CLAWSON is professor of political science, director of the Center for International and Comparative Programs, and associate director of research at The Lyman L. Lemnitzer Center for NATO Studies at Kent State University. His doctorate is from UCLA, he is a graduate of the Defense Language Institute in Russian, and has been an exchange scholar at the State University of Moscow. He has traveled widely throughout Eastern Europe and the USSR. Author of articles on Soviet-sponsored international organizations as well as domestic politics, he is coeditor of *NATO After Thirty Years* (1981) and editor of *The Warsaw Pact: Political Purpose & Military Means* (1982).

THOMAS H. ETZOLD is professor of strategy at the U.S. Naval War College. He taught at both Yale University and Miami University (Ohio) before joining the faculty of the War College in 1974. He is the author of *The Conduct of American Foreign Relations: The Other Side of Diplomacy* (1977) and has coauthored, edited, and coedited other volumes on American foreign affairs. His latest work is *Defense or Delusion? America's Military in the 1980s* (1982).

MARIANO GABRIELE is professor of contemporary history at the University of Rome and also director general of the Italian Ministry of the Budget and Economic Programs. A former NATO research fellow, his principal publications include *La politica Navale Italiana, Le convenzioni navali della triplice*, and *Operazione C 3: Malta.* He is currently working on an analysis of the Mediterranean area from World War II to the present.

LAWRENCE S. KAPLAN is university professor of history and director of The Lyman L. Lemnitzer Center for NATO Studies at Kent State University. Prior to coming to Kent State in 1954 he was with the Historical Office, Office of the Secretary of Defense. He has written numerous articles, monographs, and books on U.S. diplomatic history and NATO affairs, including *A Community of Interests: NATO and the Military Assistance Program, 1948–1951* (1980). He is editor of the center's recent publication *NATO After Thirty Years* (1981), coeditor of *The Warsaw Pact: Political Purpose & Military Means* (1982), and author of *The United States and NATO: The Formative Years* (1984).

BRUCE R. KUNIHOLM is a member of the History and Public Policy Studies departments at Duke University where he received his doctorate in 1976. He has written widely in the field of contemporary Middle Eastern affairs and won the Stuart Bernath Memorial Prize for his book *The Origins of the Cold War in the Middle East: Great Power Conflict and Diplomacy in Iran, Turkey, and Greece* (1980).

ALBERTO LI GOBBI retired in 1977 after serving as Italian military representative on the NATO Military Committee and as commanding general of the Allied Forces Southern Europe. General Li Gobbi is currently president of the Center for Defence Studies at the University of Genoa and military expert for the *Giornale Nuovo*.

RAIMONDO LURAGHI holds the chair in American history at the University of Genoa and is director of its Center for Defence Studies as well as director of graduate studies in History of the Americas. He received his Ph.D. from the University of Rome in 1958. His publications include *Storia della Guerra Civile Americana* (1966); he is currently preparing an edition of the military works of Raimondo Montecuccoli and a study of European strategy for the next generation.

ANDRÉ MARTEL is professor of history, former president of the University of Montpellier, and is now associate director of the French Institute of Comparative Military History. A member of the Scientific Council of the Foundation for National Defense Studies, he has published works on North Africa as well as on French defense institutions. He is currently working on a project focusing on French military history from 1940 to the present.

ALBANO NOGUEIRA is a retired diplomat and now visiting professor of international relations at Minho University, Portugal. He was the Portuguese permanent representative to the European Economic Community (1964–70) and to NATO (1970–74), ambassador to London (1974–76), and secretary general of the Portuguese Foreign Ministry (1976–78).

JOAQUIN OLTRA received his doctorate in Spanish law from the University of Valencia, Spain. For the last twelve years he has been professor of English and American institutions at the Autonomous University of Barcelona, faculty of letters. He is the author of *La influencia norteamericana en la Constitución española de 1869* and of several studies on the presence of American ideas in the works of nineteenth-century Spanish politicians and writers.

S. VICTOR PAPACOSMA is associate professor of history and administrative coordinator of the Center for International and Comparative Programs, Kent State University; he has taught Balkan history at Kent State since 1969. An Indiana Ph.D., he has published *The Military in Greek Politics: The 1909 Coup d'Etat* (1977) and is currently completing a study on the accession of Greece and Turkey to NATO in 1952.

STEVAN K. PAVLOWITCH teaches the history of the Balkans from Islam to communism at the University of Southampton, the United Kingdom; his special field of interest is Yugoslavia during World War II. He took degrees in history from Paris (Sorbonne) and London (King's College and School of Slavonic and East European Studies). He is the author of a volume on Yugoslavia in the series *Nations of the Modern World*, as well as of several other books and numerous articles on that country's history and contemporary affairs. Currently he is preparing a book entitled "Unconventional Perceptions of Yugoslavia, 1940–45."

ANTONIO PELLICCIA is a graduate of the Italian Air Academy. He has been a fighter pilot and is a member of the Air Training Advisory Group (SHAPE-PARIS), president of IV Working Group Trimilitary Committee, a member of the International NATO Team for the Evaluation of NATO Training Aircraft, deputy commandant of the NATO Defense College, and is currently also a member of the Center for Defence Studies at the University of Genoa.

WILLIAM H. ROWDEN is an alumnus of the U.S. Naval Academy and was commander of the U.S. Sixth Fleet from 1981 to 1983, after service as deputy chief and chief of Naval Operations Surface Warfare from 1979 to 1981. He is currently vice admiral, U.S. Navy, and commander, Military Sealift Command.

MARK R. RUBIN is assistant professor of French, associate director of the Center for International and Comparative Programs, and administrative coordinator of The Lyman L. Lemnitzer Center for NATO Studies at Kent State University. His doctorate is from Princeton University. He served as an administrative officer in the U.S. Army and has traveled extensively in Western Europe where he is a frequent field director of Kent State's Geneva Semester Program. He has a special interest in the field of French and Swiss affairs.

E. TIMOTHY SMITH is a member of the History Department at Wesleyan College, Macon, Georgia. He received his doctorate at Kent State University in 1982 and is currently preparing for publication his dissertation "The United States, Italy, and NATO: American Policy Toward Italy, 1948–1952." Part of his dissertation was printed in the spring 1983 issue of *Diplomatic History*.

Abbreviations

AAW	antiair warfare
ASROC	antisubmarine rocket
ASW	antisubmarine warfare
AWACS	airborne warning and control system
CINCSOUTH	Commander in Chief Allied Forces Southern Europe
COMEDEAST	Commander Eastern Mediterranean
COMEDNOREAST	Commander Northeast Mediterranean
DECA	Defense and Economic Cooperation Agreement
DIC	defense identification code
ETA	Euzkadi ta Azkatasuna (political party, Spain)
FEBA	forward edge of battle area
FLN	National Liberation Front (Algeria)
FLOT	forward line of troops
IBERLANT	Iberian Atlantic Area
IBRD	International Bank for Reconstruction and Development
IMET	international military education and training
IMF	International Monetary Fund
LAW	light antitank weapon
OECD	Organization for Economic Cooperation and Development
PASOK	Panhellenic Socialist Movement
PLO	Palestine Liberation Organization
SACEUR	Supreme Allied Commander Europe
SACLANT	Supreme Allied Commander Atlantic
SHAPE	Supreme Headquarters Allied Powers Europe
SWAPO	South West Africa People's Organization
TOW	tube-launched, optically tracked, wire-guided missile
UNIFIL	United Nations Interim Force in Lebanon

EAST-WEST COMPETITION IN THE MEDITERRANEAN

NATO and the Mediterranean Powers in Historical Perspective

LAWRENCE S. KAPLAN and ROBERT W. CLAWSON

Any strategic assessment of the Mediterranean basin in 1983 seemingly should identify it as a mare nostrum for NATO, as it was for Rome two millennia ago and for Great Britain during the last century. Its principal entryways are controlled by members of the alliance: Spain and the United Kingdom at the Atlantic and Turkey at the Black Sea. A third outlet at the Suez Canal is in the possession of friendly, neutral Egypt. The importance of the Mediterranean Sea to the alliance is undeniable: five of its members border its northern, western, or eastern shores, and vital oil resources from the Middle East are transported to Europe through its waters. Its military significance to NATO has been underscored by the establishment and maintenance of a southern European command headquartered in Naples—one of three major commands directly subordinate to the Supreme Allied Commander Europe (SACEUR)—and by the U.S. Sixth Fleet headquartered in Gaeta. American military installations extending from Spain to Turkey are also in service to the alliance.

The American presence alone is an impressive illustration of the Mediterranean's significance to NATO. A U.S. House of Representatives subcommittee report issued in 1977 listed 199 NATO military facilities there.[1] Since the beginnings of the Cold War in Greece and Turkey, the centerpiece of American operations has been the Sixth Fleet, a naval entity almost by itself for a generation. Normally, the fleet consists of forty to forty-five ships, including two aircraft carriers with seventy-five to ninety-five planes and one helicopter carrier with an 1,800-man assault force joined to the core units.[2] These forces, added to those of the Commander of the Allied Forces South, form an impressive military and naval power charged with formidable responsibilities.

However, any claim that the Mediterranean is a NATO lake, while not necessarily completely false, requires a degree of optimism difficult to sustain.

---

[1]U.S., Congress, House, Subcommittee on Europe and Middle East of the Committee on International Relations by the Foreign Affairs and National Defense Division, *United States Military Installations and Objectives in the Mediterranean*, 95th Cong., 1st sess., 27 March 1977, p. 8.

[2]John M. Collins, *U.S.-Soviet Military Capabilities: Concepts and Capabilities, 1960–1980*. (New York: McGraw-Hill, 1980), pp. 336–39, and tables on pp. 552–53.

Such a notion has been rendered increasingly less credible by historical experience over the past generation, by severe political and military limitations imposed by divisions within the alliance, by the hostility of some of the nonaligned nations on the southern rim of the Mediterranean basin, and by a quantitative increase in Soviet naval and air power in the area.

ORIGINS OF THE NATO CONCEPT OF THE MEDITERRANEAN: THE WEST

No broad strategic conceptualization surfaced in the seminal years of NATO. In fact, the Mediterranean existed as an almost alien and perhaps unnecessary adjunct, a kind of embarrassment to the alliance. Political turmoil in Greece and Turkey may have been the inspiration for the Truman Doctrine, but the solutions to their problems in 1947 and 1949 appeared to have been the result of bilateral efforts by the United States to defeat Communist insurgency and subversion in the former and to build up military strength in the latter. There was appreciation by the future NATO partners for America's Truman Doctrine actions, but it was limited and qualified. When Britain turned over the problems of Greece or Palestine to America for a solution, the perceived hyperboles of the doctrine were not what they had in mind. Yet in Greece and Turkey it was the Truman Doctrine that brought the Sixth Fleet into the eastern Mediterranean, initially as a gesture to the Greeks and Turks. It also laid the groundwork for military and economic aid to Western Europe.[3]

To the members of the Brussels Pact, with whom the United States negotiated for the North Atlantic Treaty in 1948 and 1949, Greece, Turkey, and the eastern Mediterranean were too far removed from the important centers of Europe, and their inclusion would have constituted a diversion of effort and a dilution of strength. Despite the concern of the United States that exclusion from NATO might seem like a sign of desertion, there was no real possibility of their admission in 1949. The most that was done was to issue a statement of solidarity within British and American pledges of support.[4]

The western Mediterranean nations were given little more encouragement. Spain was anathema as a Fascist country. Italy was suspect, partly because of the Western Union's concern that Italy would take too large a share of military aid, and to a lesser extent because the "Atlantic" element in NATO would be compromised. That Italy was included was the work of the French, who saw an opportunity to tie Italian membership to an Algeria

---

[3]P. A. Dur, "The Sixth Fleet: Search for Consensus," *Proceedings of U.S. Naval Institute* 100, no. 4 (June 1974): 20.

[4]Secretary of state to heads of missions in Greece, Turkey, and Iran, 2 April 1949, U.S., Department of State, *Papers Relating to the Foreign Relations of the United States, 1949* (Washington, DC: Government Printing Office, 1975), 4:270–71 (hereafter cited as *FRUS*).

incorporated as an integral part of France.[5] It was also a product of lobbying on the part of U.S. diplomatists like John D. Hickerson who wanted Italy inside NATO as a means of containing the powerful Communist Party.[6]

In retrospect, the Mediterranean in early NATO history appears bifurcated with the eastern and totally excluded from its provenance, while the western portion was grudgingly admitted to membership. That membership, however, seemed without any clear-cut function. Iberia was represented by a Portugal separated from the Mediterranean by Spain and valued primarily for its Azores connection; Algeria was present simply as three departments of France rather than for any strategic function; and Italy was included only as a member of the Southern Europe-Western Mediterranean Planning Group, along with France and the United Kingdom. There is little evidence that a major Mediterranean role for NATO was envisaged during this early period.

Change came dramatically with the onset of the Korean War. Under the impact of the conflict in Asia, the United States and the European allies initially pictured the Communist scenario for the divided Korean peninsula as a likely one for divided Germany. If international communism could ignite a limited war of conquest in Asia, it might well do the same in Europe. Hence, NATO was converted from a treaty of alliance, in which deterrence was expected to derive from the act of unity (Article 5), into a military organization that brought into play the mutual military assistance provided for in Article 3.

To build a massive defense structure required greatly increased American military aid and a major change in the structure of the organization. The regional planning groups, with their reflective and leisurely pace of 1949 and early 1950, were replaced with supreme allied commands, the most important serving in Paris under SACEUR. Inevitably, NATO's views on Greece and Turkey becoming members, Spain as a base, and even Yugoslavia, changed drastically.

GREECE, TURKEY, AND YUGOSLAVIA: SHIFT TOWARD THE EAST

The first reaction concerned Greece and Turkey, neither of which had given up their campaign for membership. For the alliance, an important aspect was the question of manpower resources. If German troops and material were

[5]Jean Chauvel, "Les puissances et la Méditerrané," *Politique étrangère* 36, no 5 (1971): 463–64.

[6]Note in particular Escott Reid, *Time of Fear and Hope: The Making of the North Atlantic Treaty, 1947–1949* (Toronto: McClelland and Stewart, 1977), pp. 210ff; E. Timothy Smith, "The United States, Italy and NATO: American Policy Toward Italy, 1948–1952" (Ph.D. diss., Kent State University, 1981), pp. 111ff.

to be utilized by NATO through the European Defense Community, so too Balkan forces would be significant. A Soviet-sparked European war probably would have a southern theater in addition to northern and central flanks; thus, there was now immediate receptivity to the renewed Greek and Turkish requests for membership. Yet the result was not an instant accession. The older complaint of northern Europe about distance, as well as competing military equipment appetites, recurred and was joined for a time by the concerns of American Joint Chiefs of Staff over excessive obligations in light of the Korean War. At the New York meeting of the North Atlantic Council in September 1950, a compromise offer of associate status was made as a first step toward NATO Mediterranean planning.[7] This was swept aside at the Lisbon meeting in February 1952; Greece and Turkey became the thirteenth and fourteenth members of the alliance. The expectation of twenty-five divisions, to help swell the fifty requested of the member nations at Lisbon, made their accession attractive enough to overcome all opposition.

Even Yugoslavia, whose subsequent nonalignment was put off for a time by what it thought were the lessons of Korea, seemed to wish for the protective cover of NATO in the early 1950s. It suspended its differences with Italy over Trieste and worked with NATO, ultimately signing the Treaty of Ankara in 1953 which created a new Balkan alliance, linking it with Greece and Turkey. The death of Stalin shortly thereafter freed Yugoslavia from its insecurity sufficiently to render moot the twenty-year alliance. Yugoslavia has kept its Western options open, even in its heyday as one of the leaders of the nonaligned movement.[8]

SPAIN: A DIFFERENT SCHEME

Franco's Spain was more eager to identify with NATO and was also more repugnant to most of the allies than was Tito's Yugoslavia. There always had been important sentiment in the United States for the incorporation of Spain into the NATO alliance, despite the dislike for Franco's regime and its role in World War II. On the floor of the Senate, champions such as Senator Pat McCarran of Nevada could be found to identify Spain's anticommunism as reason enough for American assistance, whatever the allies might think. At the same time, the Joint Chiefs of Staff recognized Spain's strategic

---

[7]Lawrence S. Kaplan, *A Community of Interests: NATO and the Military Assistance Program, 1948–1951* (Washington, DC: Office of Secretary of Defense, 1980), pp. 118–19. The United Kingdom's ambassador to Belgrade reported in December 1951 that Tito was ready to "assure Greece and Turkey of defence aid" if it would be reciprocated, but he was reluctant to commit himself openly. Sir Ivo Mallet to Foreign Office, 1 December 1951, no. 144, RY 1077/1-95497, London, Public Record Office.

[8]See Stevan K. Pavlowitch, "The Grey Zone on NATO's Balkan Flank," *Survey* 25 (Summer 1980): 20.

importance, even as it understood the political obstacles to its membership in NATO.[9] The Korean War deepened American interest in Spain, although only Salazar's Portugal openly championed Franco's cause. The consequence was a backdoor relationship in which the United States, propelled by the McCarran Rider to the General Appropriations Act of 1950, at first provided for funds and then for a military agreement for U.S. bases in 1953. Air and naval facilities at Rota converted Spain into an unofficial adjunct of NATO in the 1950s, a situation that persisted into the following decades as the *Polaris* submarine made that base even more important. Formal diplomatic relations were established between Spain and the United States in an exchange of ambassadors at the end of the 1950s, but it required the death of Franco before Spain could join the alliance. By that time resistance came far more from the Spanish Left than from northern Europe.

BRITISH FORCES AND FRENCH ROLES

Actually, events seem to have dictated a Mediterranean venue for NATO as early as the establishment of Supreme Headquarters Allied Powers Europe (SHAPE) and of the political headquarters in Paris. There are charts to prove it. In Lord Ismay's history of NATO, the Allied Command Europe as of 1954 embraced four subordinate commands at Supreme Headquarters. Two of them related to the Mediterranean responsibilities of NATO. Allied Forces Southern Europe, headquartered in Naples, included Allied Land Forces Southeastern Europe at Izmir, Turkey; Allied Land Forces Southern Europe at Verona, Italy; as well as Allied Air Forces Southern Europe and Naval Striking and Support Forces, both at Naples. Additionally, a second major command, Allied Forces Mediterranean, was then headquartered at Malta with an even larger number of subcommands under its aegis, ranging from the western Mediterranean—Gibraltar and Algiers through the central area at Naples—to the eastern Mediterranean area at Athens, northeastern Mediterranean area at Ankara, and finally the southeastern Mediterranean area located at Malta. The listing is impressive and suggestive. From virtual exclusion, due to the dominance of the Brussels Pact powers in the early days of NATO, the Mediterranean appears to have assumed at least equality with the other areas. The central and northern sectors, previously almost the sole concern for NATO, were somewhat diminished by the attention granted to the southern portion.

Any neglect of the other sectors was more apparent than real. The proliferation of commands in the Mediterranean could and did reflect less the primacy of their roles than the conflict and confusion over their missions. Ismay hinted as much when he laconically observed that "the problem of

[9] Memorandum, Joint Chiefs of Staff to secretary of defense, 5 January 1949, "North Atlantic Pact," *FRUS, 1949*, 4:13.

command in the southern area was more [than those in the other sectors] difficult to resolve."[10] Despite such sweeping and comprehensive titles as "southern Europe" and "Mediterranean," there was little of substance behind them; the very names themselves reflected redundancy and wasted effort. The post-Korean War reorganization of NATO ran afoul of national rivalries which included American, British, and French ambitions in the area. No single comprehensive and coherent Mediterranean policy could have possibly emerged from that competitive tangle. The result was the absence of any Mediterranean policy beyond its being the conventional "right flank" of NATO, as it was optimistically identified from General Dwight D. Eisenhower's new Paris headquarters in the summer of 1951.[11]

Difficulties had emerged initially from British opposition to U.S. dominance in any new Mediterranean Command. The United Kingdom already had been excluded from what the British had perceived to be a natural leadership role in the Atlantic Command in Norfolk and were not much appeased by the establishment of a special Allied Command Channel under British responsibility. Moreover, their conception of the Mediterranean, particularly the eastern Mediterranean, had differed from that of the United States. Turkey, in Britain's estimation, was more properly a part of a Middle Eastern Command under British leadership. That idea had been discussed in 1952 but not acted upon until the establishment of the Baghdad Pact later in the decade.

British memories of past glory were made more painful by the massive presence of the U.S. Sixth Fleet, which would remain a separate entity no matter what form of organization the NATO commands might finally take. The trouble, as Ismay diplomatically put it, was that "the special position of the British Naval Force which had for so long wielded control of the Mediterranean could not continue."[12]

Duplication resulted. The Allied Command Southern Europe was hastily established in June 1951 with an American admiral serving as commander in chief, an Italian general in charge of land forces, an American general to head the air forces, and an American admiral to command the Sixth Fleet. After Greece and Turkey joined in July 1952, southern Europe was given additional responsibility for a southeastern command with headquarters at Izmir and an advance post at Thessaloniki. In March 1953 a Mediterranean Command, Allied Forces Mediterranean, came into existence with Admiral of the Fleet Earl Mountbatten of Burma as its first commander. While the national forces subordinate to him would be in six separate areas, in time of war Mountbatten would have the responsibility for the security of the sea lines of communications throughout the Mediterranean.[13] Each of these numerous commands was the result of various compromises to appease national

[10]Lord Ismay, *NATO: The First Five Years, 1949–1954* (Paris: NATO, 1955), p. 73.

[11]Hamilton Fish Armstrong, "Eisenhower's Right Flank," *Foreign Affairs* 29, no. 4 (July 1951).

[12]Ismay, *NATO*, p. 73.

[13]Ibid.

sensibilities. This is the principal explanation for the somewhat bizarre alignment of subordinate commands in the southern sector of SHAPE.

The lengthy delay between the NATO Council decision to establish an overall British-led Mediterranean Command in 1952 and its implementation over one year later was essentially because of conflict between the American insistence that the officer designated report to Commander in Chief Allied Forces Southern Europe (CINCSOUTH) in Naples and the British position that he report directly to the Standing Group in Washington. The essence of the eventual compromise was that the British admiral should be subordinate to SACEUR in Paris rather than to either CINCSOUTH or the Standing Group. This compromise was made against the advice of the American Joint Chiefs of Staff. The United States was deferring to a longstanding British demand for a Middle East Command, in which Turkey might provide a kind of link to Arab states in the area.[14]

Uncomfortable as Anglo-American relations were over the Mediterranean, the French reaction to what they saw as an Anglo-Saxon condominium was even more troublesome. There was a passion behind the French position that was absent in the British and American differences. Like Britain, France remembered its past, but it found no special relationship with America to compensate for its losses. And, unlike the United States or the United Kingdom, France was a Mediterranean power with its sovereignty extending to both the northern and southern shores of the sea. For France the Anglo-Saxon Mediterranean policy dramatized what was wrong with military integration: it excluded France. France secretly conditioned its approval of Greece and Turkey's joining NATO at the time of the Ottawa meeting of the North Atlantic Council; the French western Mediterranean naval commander would have to be placed directly under U.S. Admiral Robert B. Carney, CINCSOUTH, rather than under a British superior. This was the best France could achieve in the face of American and British power.[15]

More than six years after the controversy over the Mediterranean Command was concluded, René Pleven, premier during a part of the period, was still angry over Mountbatten's role in Malta, where France's naval capabilities seemed equal to those of the British.[16] It was clearly as appropriate for France to have a central role in the Mediterranean as it was to have such a responsibility in Atlantic Europe. When Algeria broke away from the French Union

[14]See memorandum by chief of staff, U.S. Army, to Joint Chiefs of Staff, "Organization of the Defense of the Middle East," 10 December 1952, JCS 1868/431, RG 218, National Archives and Records Service. France was equally concerned with recognizing "the Middle East as a theater of operation second only to Western Europe," even if it would be nominally outside NATO's jurisdiction. See memorandum for General Omar Bradley by Vice Admiral Jerauld Wright, U.S. deputy representative to the surgeon general, "Mediterranean Command Discussion of in Standing Group," 17 April 1951, CJCS 092.2, RG 218, ibid.

[15]Memorandum for the president by General Bradley, 18 October 1951, CJCS 092.2, RG 218, ibid.

[16]René Pleven, "France in the Atlantic Community," *Foreign Affairs* 38, no. 4 (October 1959): 27–29.

in 1962, no diminution of involvement should have resulted. Special ties with the Maghreb endured, and a new function was added as France disengaged from a de facto Israeli alliance and sought increasingly to identify itself as a protector of the Arab Middle East.

In certain respects the events of the mid-1950s served as a watershed for French—and British—concern for the eastern Mediterranean. First, the ill-fated Anglo-French Suez intervention heralded the decline of both European naval powers in the Mediterranean as well as their replacement initially by an unchallenged American presence. Ultimately, there was to be what Jean Chauvel, former secretary general of the French foreign ministry, called "a double domination" of superpowers as the Soviets increased their attention to the seas.[17] Second, the eastern Mediterranean reassumed its primacy as a focus of attention as the Cyprus problem manifested itself in 1955 and as the Arab-Israeli conflict assumed crisis proportions during the next year.

A generation later in the 1980s the two issues remain paramount for NATO. Greek and Turk, Israeli and Arab, seem no closer to the resolution of their differences in 1983 than they had in the 1950s, which means that the Mediterranean remains a potential flash point for both Soviet and American interests.

In the de Gaulle era—and even after de Gaulle—French thinking about the Mediterranean turned to the idea of excluding the sea from the Cold War, which in effect meant reducing, if not eliminating, the presence of the superpowers in the area. To dramatize the issue, de Gaulle removed the French navy from the Mediterranean in 1959. This was to have been a signal to the United States that France would have no part of any integrated naval activity in the area and to the Soviet Union that, if NATO and France reduced or removed their forces, the pretext for increasing the number of Soviet warships also would be removed.

France looked for support for its position from Spain and Italy but without much success. For a time the idea of *Europe des patries* appeared to be a Spanish option for Europeanization during its long, formal exclusion from NATO, based on the possibility of a Mediterranean pact on the Gaullist model. This idea eventually foundered in Italy's continuing attachment to NATO and U.S. leadership. It could not survive Spain's interest in providing American bases or France's own ambivalence toward NATO, which manifested itself in the Mediterranean just as it did elsewhere.

De Gaulle eventually took his nation out of the organization but not out of the alliance. A recognition that even the organization might be needed in a crisis has persisted, and France has accommodated itself to that possibility. French naval participation in NATO exercises and its continuing contacts with Maritime Air Forces Mediterranean (MARAIRMED) were characteristic of cooperation in the 1960s as well as the post-de Gaulle 1970s. France under

[17]Chauvel, "Les puissances et la Méditerranée," p. 464.

François Mitterrand in the 1980s is more likely to strengthen than weaken those ties.

French Mediterranean policy under de Gaulle stressed regionalism over globalism; he could maneuver among the great powers to the degree that the western Mediterranean states accepted French leadership. Arabs from Morocco to Libya could be wooed, and even a Fascist Spain might be brought into the associations. To enlarge on the prospect of "*la Méditerranée aux Méditerranéens,*" France provided aid to Libya and adopted an anti-Israel foreign policy. As former Foreign Minister Maurice Schumann observed, "there is no French policy on arms contracts, there is a Mediterranean policy for France which . . . has only one aim to prevent . . . the western Mediterranean . . . and if possible . . . the whole of the Mediterranean basin from being a theater of the Cold War."[18]

The British after Suez were even less successful than the French. As they removed themselves from imperial responsibilities, the British role dwindled to that of an honest broker: an experienced hand that could speak for Europe more objectively than the United States. Such was its voice in the European Economic Community, as represented by Lord Carrington at the end of the 1970s, proposing a solution to the Arab-Israeli problems separate from that of the United States. NATO would be served by a separate British contribution, not as a junior partner in an Anglo-American combination. The disappearance of the Mediterranean Command—the British prize of 1953—at the end of the 1960s was eloquent testimony to the demise of British ambitions.[19]

The Sixth Fleet and the CINCSOUTH remain paramount. Unlike France, Britain does not have a regional policy of neutralism. The United Kingdom is essentially an observer of the superpower conflict in the Arab-Israeli struggle, no longer master of Suez. Its Mediterranean position is also challenged by increasing Spanish pressure for the return of Gibraltar. This is not to suggest that there has been no role for the British fleet after Suez. Gibraltar was still seen as a bastion of empire in the NATO service, and Britain still had a legal position on Cyprus as well as on two bases. And British Mediterranean naval forces, though certainly diminished, have by no means disappeared in the 1980s.

The U.S. role as the representative of NATO, and then as a force independent of NATO, has been the norm for almost a generation. Dating from the U.S. opposition to the Anglo-French invasion of Egypt in 1956 and the brief excursion into Beirut in 1958, the Mediterranean has been largely an American theater; in the late 1950s and early 1960s almost exclusively

[18]Quoted in Edward A. Kolodziej, "French Mediterranean Policy: The Politics of Weakness," *International Affairs* 47, no. 3 (July 1971): 509.

[19]Headquarters Allied Forces Mediterranean was deactivated as of 5 June 1967. SHAPE General Order 10, 6 June 1967.

so. Consequently, Soviet naval power has been met primarily by an American rather than a NATO response.

THE SOVIET CHALLENGE

The inhibiting presence of the Soviet Union, particularly its naval presence since the mid-1960s, has been the most obvious and visible of the problems requiring NATO's attention to the Mediterranean. Soviet, and earliest tsarist Russian, interest in egress from the Black Sea has been a relatively constant factor for at least two centuries. As the Russian Empire emerged as an expansive force in the eighteenth century, access to open water became an important element of tsarist foreign policy. A central facet of the long Russian-Turkish conflict was first over Russian expansion to the shores of the Black Sea and Caucasus Mountains, and then over access to the Mediterranean through the Dardanelles. Soviet victory in World War II revived and intensified older Russian Mediterranean concerns and opened new opportunities through an armed Communist insurgency movement in Greece. Further to the West the Soviet Union claimed a role in the Italian peace settlement and sought unsuccessfully, although not too vigorously, to be a trustee of Italy's Libyan colony. To the south and east the USSR supported Arab rights against British and French colonialism; it even voted in favor of the creation of Israel in 1948 to force Britain out of Palestine.

Over the next generation the Soviet presence waxed and waned but never disappeared. Soviet influence may have been checked, for example, in Greece and Turkey, in part by American actions under the Truman Doctrine and later by membership in NATO. The USSR's prospects were further reduced by rebuff from Tito's Yugoslavia in 1948 and more resolutely from 1953 to 1955 when a fragile Balkan Pact linked Yugoslavia to Greece and Turkey. But the Soviets returned with renewed vigor through their dealings with Egypt and then Syria from the mid-1950s onward and by their interaction with the radical Arab nations, generally in their role as champions of anti-Western "national liberation" movements.

Part of the potential Soviet threat was always military. As SHAPE came into being in 1951, the idea of a European Command of three parts—northern, central, and southern—emphasized the importance of the southern flank in NATO planning. Greece and Turkey and, to an extent, Yugoslavia were in a geographic position to discourage or contain any Soviet thrust from the Caucasus or from Romania or Bulgaria that could threaten Yugoslavia and Italy. And a successful Soviet offensive, political or military, could undermine whatever stability might have been achieved in the northern and central sectors. As Hamilton Fish Armstrong noted in 1951 when the prospect of Greek and Turkish membership seemed likely, their armies totaled 750,000 men if added to a Balkan alignment that included Yugoslavia.[20] Since then the defense of the eastern Mediterranean against Warsaw Pact armies has remained a continuing responsibility for the Allied Forces South Command or Mediterranean Command.

It has been Soviet naval penetration of the Mediterranean, however, which has merited most of NATO's attention there. If Greek and Turkish membership in NATO, and Yugoslavia's linkage through a Balkan Pact, were interpreted by the Soviets as a threat, then Secretary of State John Foster Dulles's initiative to fashion the Baghdad Pact—a Middle Eastern extension of NATO—provided an even greater challenge. Yet at the same time, NATO's weaknesses were being exposed.

The violent manifestation of Greek nationalism in Cyprus in 1955, the beginning of a new generation-long crisis between Greece and Turkey, combined with Nasser's welcome of Soviet support in the contest over the Suez Canal provided the Soviet Union with an opportunity to secure safe havens for its own naval forces as it leapfrogged over the feeble Baghdad Pact and even over NATO military lines themselves. Still, despite periodic Soviet threats—from military support of Egypt in 1956 to demands that Turkey dismantle the *Jupiter* missiles in 1962—the actual presence of a Soviet navy in the Mediterranean was for many years not substantial. In fact, when Albania defected from the Soviet orbit and roughly turned the Soviet Union out of its submarine base at Valona in 1958, there were for a time no Soviet vessels regularly stationed in the Mediterranean.

Changes came dramatically in the 1960s as favorable circumstances of all kinds seemed to invite Soviet opportunism. A prime occasion was provided by the crisis in Cyprus which flared again in 1963. Additionally, the Soviets were encouraged by the withdrawal of the French navy from the Mediterranean Command of NATO, the Six-Day War between the Arabs and Israel, and the steady reduction of Western influence in the North African littoral, particularly in Libya and Egypt. The subsequent buildup of Soviet naval forces from twelve units in 1966 to seventy in 1970, with major port facilities available in Alexandria and Port Said, took advantage of these Western vulnerabilities.

The Soviet Union made no secret of its position; Leonid Brezhnev, in a declaration at Karlovy Vary as early as April 1967, declared the Soviet Union to be a Mediterranean power and its antagonist to be the American Sixth Fleet.[21] Brezhnev's triumphant announcement apparently coincided with

---

[20]Armstrong, "Eisenhower's Right Flank," p. 652.

[21]*Rech glavny delegatsii KPSS tovarishcha L. I. Brezhnev*" [speech by the CPSU delegation chief L. I. Brezhnev], in *Pravda*, 25 April 1967. This important foreign policy statement pinpointing, among many other things, a new intiative in the eastern Mediterranean, was made at the carefully prepared Conference of European Communist and Workers' Parties in Czechoslovakia. It was presented more than six months before the Six-Day War and actually before specific events had begun to indicate the possibility of a new round of Arab-Israeli combat. The

Admiral Sergei Gorshkov's emerging view of the Soviet navy as a foreign policy instrument with worldwide applications.[22]

Two major reports to NATO underscored the impact of Brezhnev's assertions; they focused on the massive military equipment aid to the Arabs which replaced its losses in full after the total destruction of the 1967 war. The report also noted the complete diplomatic support for the Arab cause in the United Nations and the quadrupling of the Soviet naval presence since 1964. By 1969 the Soviet Mediterranean contingent was larger than that of all the Mediterranean units of Western Europe put together. Even if its forces were still inferior to the fighting power of the Sixth Fleet, the Soviet Union had become a much more credible threat to Western shipping, the supply of oil, and to Western access to petroleum resources in the Middle East.[23]

If the Mediterranean had become an American zone of influence for a time in the mid-1950s, by the late 1960s it was no longer so.[24] Today the Soviet Union can deploy naval and supporting naval air capabilities comparable to NATO's, maybe even more. In a somber 1982 review, NATO Secretary General Joseph Luns observed that the Syrian-Israeli conflict of 1981 over increased Syrian penetration of Lebanon had resulted in the presence of a large Soviet naval task force led by helicopter-carrying aviation ships. The group cruised for two days off the Syrian coast. After the Israeli raid on Iraq's nuclear facility, the Soviets held major Mediterranean naval exercises, including an amphibious landing on the Syrian coast, their first use of naval infantry for power projection in the Middle East.[25]

Not that NATO has completely failed to respond over the years. Following the major reports in June 1968 at the Reykjavik meetings, the Defense

---

only major event of recent note had been the military coup d'état in Greece, against which all of the delegates protested. Thus the new Soviet policy initiative into the Mediterranean, which is often thought of as having been triggered by the Six-Day War, seems to have slightly earlier origins.

[22]Admiral of the fleet and commander in chief of the Soviet navy, Sergei Gorshkov, has been in charge of Soviet naval operations since the Khrushchev years. His articles and books have attracted a great deal of attention abroad. Whether or not the modern Soviet fleet was constructed and molded primarily for power projection or for specific war-fighting tasks that only collaterally lent muscle to a Soviet diplomacy of force is a question of complex controversy in the West. A comprehensive argument in favor of the former interpretation is provided in Bradford Dismukes and James M. McConnell, eds., *Soviet Naval Diplomacy* (New York: Pergamon, 1979). For a detailed discussion of the Gorshkov publications, see essays on the subject in Paul J. Murphy, ed., *Naval Power in Soviet Policy* (Washington, DC: Government Printing Office, 1978).

[23]Reports of Western European Union, Committee on Defense Questions and Armaments and of General Affairs Committee, in "Soviet Threat in the Mediterranean, *NATO Letter* 17, no. 6 (June 1969): 2.

[24]Alvin J. Cottrell and Admiral Thomas H. Morrer, "Projecting American Military Power in the Mediterranean and Western Pacific," *Military Review* 18, no. 2 (June 1978): 25.

[25]Joseph M. A. H. Luns, "Political-Military Implications of Soviet Naval Expansion," *NATO Review* 30, no. 1 (February 1982): 4–5.

Planning Committee established MARAIRMED to coordinate air surveillance.[26] A new NATO command was established for it in Naples, and the Sixth Fleet remains stronger than in the previous generation. The joining of Spain to NATO underscores continuity of control over the Strait of Gibraltar, and the return of Turkey to regular participation in NATO assures political and military possession of the Dardanelles. As might be expected, the Soviets have suffered their own political failures in the region, some of which have had long-term consequence.

How impressive, however, is the NATO presence today? The best U.S. leadership seems to offer, as Stefano Silvestri put it a decade ago, has been a "continuation of the colonial policy" of Great Britain. The failure of Britain and France to maintain their traditional control of the Mediterranean made the United States their heir. In attempting to keep the peace in the region on behalf of their own interests, Americans hardly seem to have gone beyond the European imperial powers of the past. In fact, the replacement of Great Britain with the United States in Greece appears to be little more than a "changing of the guard." The American role, which is essentially NATO's, seems passive, if not negative, seeking only to maintain the status quo.[27]

THE SIXTH FLEET

There is a widespread tendency to see a long line of continuing Soviet successes at NATO and America's expense. Nevertheless, the record suggests otherwise. On the Cyprus issue alone the USSR has gained less than that conflict might have provided. Of all the leaders in the area, Archbishop Makarios was the most useful to the Soviets in light of his clear identification with nonaligned politics. His departure produced no successor with his kind of neutralist inclinations or connections, nor does the present leadership of the Greek or Turkish Cypriot communities seem to support a separatism that would serve Soviet ambitions. Even on the mainland itself, Andreas Papandreou's widely noted election campaign rhetoric against NATO and American bases was not followed by Greek abandonment of NATO or repudiation of an American presence. Similarly, Turkish anger—expressed in the closing of bases to American use and in a limited détente with its Soviet neighbor—had its bounds. The current military regime, while exercised over Greek extension of claims in the Aegean, seeks a way to maintain its NATO connections.

Nor has the Soviet Union been particularly successful elsewhere. In the Middle East the volatility of the area has made it clear to Moscow that there

[26]"Soviet Threat in the Mediterranean," *NATO Letter* 17, no. 1 (January 1969): 9.

[27]Stefano Silvestri, "NATO and the Mediterranean Situation," in *NATO and Security in the Seventies*, ed., Frans A. M. Altung von Geusau (Leyden: A. W. Sijthoff, 1971), p. 45.

are severe limits to the value of reliability of Arab connections. At the same time, NATO still has better logistical staying power with its control of choke points in Turkey and Gibraltar and its formidable Mediterranean naval, air, and military bases, as opposed to the USSR's anchorages at long distance from its bases in the Black Sea. In brief, the Sixth Fleet with the support of MARAIRMED, the Italian positions in the Otranto Straits area, and a revived French presence, as well as Portuguese and now Spanish naval strength in the West, reduces the dangers of Soviet naval activity except in low-risk situations. While a question may be raised over which superpower has the most vital stake in the Mediterranean, the Sixth Fleet remains the cutting edge of NATO's Mediterranean force in the 1980s and continues to be an agent for stability.[28] On the other hand, the balance today is delicate; a Greek or Turkish departure from NATO, a closer Arab-Soviet alliance, an aggravated Spanish-British conflict over Gibraltar, or the removal or reduction of the Sixth Fleet could fundamentally jeopardize the alliance and the stability of the region.

Even after so many years the Mediterranean sector of NATO still lacks coherence, vitality, and meaning. This confusion is expressed mostly in negative terms. The most appealing to many, yet the most divisive for NATO, was articulated in the slogan "the Mediterranean for Mediterraneans," implying the exclusion of superpower rivalry from the region. If this could truly be achieved, it might indeed solve such major problems as Soviet opportunism in the eastern Mediterranean. But the prospect of a neutralized Mediterranean is not only chimerical; if it could be realized, it also would eliminate the stabilizing role of NATO strength as represented most dramatically by the American Sixth Fleet. Ideally, a solution might be to call on the NATO nations themselves to find what André Beaufre once termed a "homeopathic solution" in place of crisis management by the Sixth Fleet.[29] This would require an unprecedented unity of effort and of understanding as well as a common political will on the part of the European powers.

If the entire Mediterranean is not a coherent concept, then at least the eastern Mediterranean from Turkey to Lebanon and Egypt is, and a volatile one at that. The danger of a major new flare-up of Arab-Israeli hostilities is always present. Greeks and Turks continue their debilitating struggle focused on Cyprus and on an ever-worsening crisis over a cluster of disputes, including

---

[28]See Curt Gasteyger, "The Super-Powers in the Mediterranean," *Atlantic Community Quarterly* 14, no. 1 (Spring 1976): 53. The author believes that "Soviet interests seem, on the whole, to be more and less subject to change than American. More permanent because the Mediterranean is obviously much more directly linked to Soviet than to American security." In contrast, see Edward N. Luttwak, "Seapower in the Mediterranean: Political Utility and Military Constraints," *The Washington Papers* 61 (Beverly Hills, CA: Sage Publications, 1979): 17–18. He claims that the Mediterranean represents a "core value" for the United States, whereas it is a peripheral area for the Soviet Union.

[29]Beaufre presented his ideas in *Batir l'avenir* (Paris: Calmann-Levy, 1967), pp. 165–68. See also Silvestri, "NATO and the Mediterranean Situation," pp. 53–55.

territorial waters, the continental shelf, and airspace over the Aegean. They are weakening the integrity of the alliance, and, more specifically, they have reduced regional military cooperation in the eastern Mediterranean far below a credible level.

Yet neither Greece nor Turkey wishes to leave the alliance. Greece wants both a guarantee for a NATO security, and Turkey wants military aid, at least $1 billion annually in U.S. assistance. Here NATO, and particularly the United States, can play a role as it seeks to limit and ultimately control the confrontation, for the sake of all parties concerned. This primary issue was recognized in a U.S. Senate report on NATO released in April 1982, which contained nothing about the Mediterranean as a whole but a great deal about Greece and Turkey. The Reagan administration, it claimed, seeks to "maintain a stable military balance in the Eastern Mediterranean without accepting the concept of any fixed-ratio while helping both Greece and Turkey to fulfill their respective NATO responsibilities."[30] What is most needed is patience, consistency, and evenhandedness in the effort. American "gestures" designed just to placate one or another domestic lobby have no place in this critical endeavor. And while there is no grand design in prospect, the possibility of maintaining stability does exist. It is a role that NATO and the Sixth Fleet have attempted to play for a generation. They must continue to do so in the immediate future, hoping that U.S. political decision makers may one day recognize the necessity of devising effective policy in the area and act accordingly.[31]

---

[30] U.S., Congress, Senate, Committee on Foreign Relations, *NATO Today: The Alliance in Evolution*, 97th Cong., 2d sess., April 1982. See chap. 7, "Implication for the U.S. and NATO," pp. 86–96.

[31] This view is sharply challenged by U.S. Navy Commander P. T. Deutermann, who believes that the Sixth Fleet should be disestablished and deployed elsewhere in NATO's service and that its functions should be replaced by allied navies supplemented from time to time with U.S.-represented national units. "Requiem for the Sixth Fleet?" *Proceedings of the U.S. Naval Institute* 108, no. 9 (September 1982): 46–49.

The Mediterranean: Environment of the Sixth Fleet

WILLIAM H. ROWDEN

The Mediterranean has played a central role in the development of Western civilization, both as a crucial highway for commerce and as a recurring battlesite. A host of famous sea battles has been waged there. For example, Salamis (480 B.C.), Actium (31 B.C.), Lepanto (1571), Sicily (1676), and The Nile (1798) were not only actions memorialized in maritime annals, but each also had a profound impact on the history of Western Europe. Not as celebrated, but equally as significant, were the long periods of peace interspersed between these landmark battles when one power or another dominated this ancient sea and exploited it for economic purposes or to project power inland from the littoral.

The traditional pattern has continued in this century. From Trafalgar (1805) until the beginning of World War I, the British Royal Navy was essentially unopposed in the Mediterranean. Even in World War I the Triple Alliance was unable to pose a serious challenge in those waters. On the other hand, Whitehall employed its preeminence at sea to blockade Turkey; to land at Gallipoli, although not to prevail there; to mount Edmund Allenby's successful Palestinian campaign; to fight at Salonika; to control the Suez Canal; and in general to prevent imperial Germany from spreading its influence southward.

The picture in World War II was considerably different. Hitler, in league with his Italian allies, temporarily wrested control of the central Mediterranean and in turn threw his panzers into North Africa with the objective of seizing the Suez Canal and driving the British out of the Near East. He came dangerously close to succeeding. Even more than General Bernard Montgomery's genius on land, it was Britain's ability to resupply its armies through the Red Sea and eventually to sever the Afrika Korps' maritime supply lines that defeated Rommel. The seesaw battle for control of the Mediterranean from 1940 to 1942 was another of those epic watershed Mediterranean campaigns.

With the defeat of the Axis in 1945, Allied forces, particularly the Royal and U.S. navies, literally dominated the oceans. When NATO drew together the nations of the Western alliance, it was recognized that the key to a coordinated defense of Italy, Greece, and Turkey was control of the Mediterranean, and that the predominant sea power of the alliance compensated for many of the southern flank's vulnerabilities on land. Facilitating this strategy was the fact that the Soviet Union's primary objectives lay elsewhere and the Mediterranean was considered a "peripheral" area. On occasion the Sixth Fleet used, or was poised to use, its superiority in the Mediterranean to influence events ashore. Witness, for instance, its role in the Greek civil war (1946–49), the Suez crisis (1956), Lebanon (1958), and the two Sinai wars (1967 and 1973). While these events illustrated an imaginative employment of seaborne forces, the fact remains that the Mediterranean Sea itself has not seen a major naval engagement for almost four decades.

This does not mean, however, that circumstances have stood still during that period. Certain fundamental changes—economic, military, and political—have been at work and, in turn, have altered the strategic environment markedly. Much of this transformation has proceeded at such a gradual pace that it either has gone unnoticed, or at least has not proved sufficiently alarming to provoke a profound reaction. Nevertheless, the Mediterranean now once again has become a critical commercial crossroad and potential confrontation zone. The impending struggle for its control, as so often in the past, could well influence the course of history.

THE MEDITERRANEAN'S ECONOMIC IMPORTANCE

It was inevitable that Europe's postwar recovery would revive the Mediterranean as a major trade artery. What was not expected or predicted was the phenomenal growth of manufactured goods in the Western world, the insatiable appetite of modern industry for raw materials, the dramatic increase of imports and exports, and the resulting pattern of barter and mutual dependence between the developed and undeveloped nations. These factors have combined to drive seaborne traffic to unimagined heights, and the level of trade of Mediterranean littoral countries has mushroomed accordingly. It now involves the loading and unloading of an estimated 1 billion tons of cargo annually. On any given day 1,500 oceangoing ships and 5,000 smaller craft ply the waters from Suez to Gibraltar; it has been asserted that this is about one-half of the ocean traffic throughout NATO's maritime area.

About 65 percent of Mediterranean trade (in dollar value) is intraregional, the exchange of goods between the Continent, North Africa, and the eastern Mediterranean countries. Western Europe dominates this regional trade as the largest market for raw materials and the principal supplier of manufactured goods. Virtually every littoral government is dependent upon seaborne imports for at least one category of vital products. Food consumes from 15 to 30 percent of the import dollar in eleven of these nations. Egypt is by far the most reliant upon external foodstuffs, while Turkey and Greece are major food exporters. Manufactured goods absorb 40 percent of Spain's import dollars, as opposed to Libya's 75 percent. The latter statistic is typical of the oil-rich countries that in the last twenty-five years have become heavily reliant on European essential goods and services. Every Mediterranean country expends from 7 to 19 percent of its import monies on ores, metals, or agricultural raw materials.

Petroleum statistics, however, best illustrate the resurging economic importance of the Mediterranean. Four regional nations—Tunisia, Algeria, Libya, and Egypt—are self-sufficient in oil. As a percentage of the import dollar Turkey expends the most, then Spain, Italy, France, and Greece. At any given moment 300 to 400 tankers carrying 25 to 30 million barrels of oil from the Persian Gulf region, North Africa, and Near Eastern pipeline outlets are transiting this body of water. One-half of the oil consumed by France, Spain, and West Germany, and all that of Italy, Switzerland, and Austria, arrives through Mediterranean ports. As a result of the upheaval in Iran, North Africa supplies close to one-half of the petroleum imported through Southern Region ports. Approximately 50 percent of the oil produced in Algeria, Tunisia, and Libya, and about two-thirds of Syria and Iraq's export petroleum, moves to Western Europe by ship. Further influencing this picture, the Suez Canal has been enlarged to accommodate tankers of 150,000 tons, and virtually all the economic factors that count—tonnages, distances, and tanker logistics—point to a reemergence of the Suez route as a primary petroleum artery that will ultimately increase some of the foregoing figures.

In essence, the health and survival of the politically diverse countries of the Mediterranean littoral have become dependent on an interlocking pattern of seaborne trade. Taken by itself, the Western industrial world's, particularly NATO's, heavy reliance on the petroleum of North Africa and the Middle East has placed the Mediterranean in a new economic and strategic light. Strategic stockpiles in the region are practically nonexistent, and many of the littoral nations have marginal international monetary reserves. Ironically, the largest oil-producing countries that do have comfortable reserves, such as Algeria and Libya, require constant supplies of food and goods to keep their "made in Europe" infrastructure functioning. Clearly, freedom to use the seas can spell the difference between economic stability or chaos.

This description is only part of the economic portrait. In the last thirty years the Mediterranean also has become an important conduit for the Warsaw Pact countries. At any moment an average of 150 Soviet merchantmen are sailing its waters. Over 60 percent of the USSR's exports and more than 50 percent of its imports, not to mention those of Bulgaria and Romania, are funneled through the Bosphorus. It is estimated that Pact ports in the Black Sea handle as much as 150 million tons of cargo, and that over 18,000 ship transits of the Turkish Straits are made yearly. This is the Soviet Union's primary link to non-Warsaw Pact client states such as Ethiopia, South Yemen, Syria, Libya, and Cuba. Unlike the West, the USSR is relatively self-sufficient in oil and mineral resources but must import significant amounts of grain. This trade, combined with the West's dependence on the Mediterranean, explains in large part the USSR's emerging political and military involvement in the area.

Unlike the early postwar years, there are now compelling Soviet strategic objectives on NATO's southern flank. The political and economic rewards of Soviet domination of the Mediterranean, or just the eastern Mediterranean, would be extremely high. As for NATO, the commercial imperatives are such that even in war the alliance could not abandon its Third World trading partners or the trade lanes of the Mediterranean. To do so would result in serious interruption of resources to the West, risk potential instabilities from North Africa to the Middle East, and present golden opportunities to the Kremlin to win support in the Third World at NATO's expense. These linkages would have to be maintained in a general conventional war by such NATO measures as production for export as well as defense and an early effort to secure the necessary sea lines of communication. A genuine disruption of these trade lanes for a period of thirty days or more could result in the economic collapse of weaker allies and some lesser developed littoral countries and have a severe impact on the entire Mediterranean economic structure.

THE SOVIET NAVAL CHALLENGE

Commercial interests and strategic objectives may cause friction between governments, but they are not sufficient in themselves to ensure a military challenge. An authentic naval rivalry requires two opponents with the capability to contest the sea in question. As emphasized earlier, the postwar Soviet navy had neither the weaponry nor the reach to dispute the Mediterranean. It was essentially a coastal defense force that had played only a peripheral part in the Soviet victory over the Nazis and enjoyed little influence in a country recently and deeply indebted to its ground forces and in a defense bureaucracy dominated by generals.

It took twenty years, an expanding global role, a skillful political admiral, and recurring demonstrations of the importance of Western sea power to change the Soviet hierarchy's attitude toward maritime affairs. But once done, the results have been dramatic. When historians look back, they may well conclude that the most remarkable military achievement of this era was the transformation in two decades of a small Russian inshore fleet into a sophisticated, blue water navy capable of global operations.

Nowhere is this more graphically illustrated than in the Mediterranean. The Soviet permanent presence commenced modestly in 1969 with only a token force of four or five ships. In the next ten years the Soviet Mediterranean squadron (the Fifth *Eskadra*) grew steadily, reaching an average daily level of over fifty combatants, submarines, and auxiliaries. Moreover, these deployment levels are augmented during periods of crisis by surface ships from the Black Sea and by submarines from the Soviet Northern fleet. The extent to which the Mediterranean *Eskadra* would be augmented preceding major hostilities is dependent upon circumstances, but it is possible that peacetime deployments would be tripled, while still leaving a formidable Soviet naval force just a short distance away in the Black Sea.

The potential Soviet maritime threat in the Mediterranean is, in sum, three-dimensional: surface, subsurface, and air. Before engaging, surface forces would very likely be organized into task groups, each of which would include long-range missile shooters and be deployed within range of NATO's principal units, particularly the carriers. Soviet naval surface-to-surface missile capability is perhaps the world's most formidable, including SSN-2s, 3Bs, 9s, and 12s. If these units were permitted an unopposed first strike, they could inflict considerable damage at the early stages of a conflict. To compensate for lack of tactical air support, such formations would be heavily armed with surface-to-air missiles capable of engaging high-performance aircraft.

Below the surface, the most threatening of the Soviet submarines are the *Echo II*s and *Julietts*, each carrying 200 nautical mile range-cruise missiles. Next are the *Charlie* class boats with sixty-mile weapons and the nuclear attack *Victor* class carrying antisubmarine missiles. Supplementing these are large numbers of torpedo-firing diesel submarines that would necessarily represent a significant proportion of any deployed force. Due to difficult hydrographic conditions in the Mediterranean and the lack of wartime experience with both nuclear-propelled and missile-firing submarines, the underwater challenge presents a host of uncertainties.

The most effective aircraft in the Black Sea naval air inventory are *Badger* and supersonic *Backfire* bombers. The latter are among the most advanced heavy bomber aircraft in the world and can be intercepted by only the most sophisticated fighters. These units each carry two 180 NM air-to-surface missiles and have the range to attack any maritime target in the Southern Region. Operating from its permanent bases, Soviet naval air power can be augmented, if necessary, from other bomber forces in the southern USSR and with additional *Backfires* from elsewhere in the Soviet Union.

This picture though is not all bright for the Soviet navy. It has only limited access to ship repair facilities and resupply installations on the littoral of the Mediterranean. Admiral Sergei Gorshkov has made an impressive effort to compensate by acquiring a variety of support ships that sustain the *Eskadra* in anchorages off Tunisia, Libya, Greece, Cyprus, and Egypt. The extensive logistic train must be protected and complicates the commanders' problems. Moreover, despite the emphasis placed on service units, the Soviets' replenishment techniques afloat continue to lag behind those of the U.S. Navy. In the event of hostilities, the lack of seaborne air cover would require surface units to rely totally on surface-to-air missile protection, and, although the Soviet missiles are sophisticated, this would still leave a chink in their armor until Soviet land force victories could bring Soviet Frontal Aviation fighter-bombers into position to roam the Mediterranean. Moreover, for reasons that are not totally discernible, Soviet units train less than their Western counterparts. This phenomenon is partially the consequence of a different basic Soviet naval philosophy that stresses equipment maintenance rather than operation. However, nothing can substitute for the actual practice of what would be wartime functions. Finally, geography plagues Gorshkov. The Mediterranean is surrounded by choke points, all of which pose dilemmas for Moscow. The most crucial are the Bosphorus and Dardanelles. That problem cannot be solved solely by naval operations. If the Fifth *Eskadra* is to be sustained or reinforced in prolonged hostilities, the Soviet navy must look to its ground forces to overrun the Straits.

While these considerations lend balance to the picture, they do little to discredit the fact that the Soviet Union is now a genuine Mediterranean naval power. In war its fleet can seriously contest the West's supremacy in the central and eastern Mediterranean and attack the sea lines of communication as far west as Gibraltar. This single development has dramatically altered its strategic equation in the Mediterranean and eroded the initial premises of NATO's southern front strategy. No longer can the coalition assume uninterrupted supply or reinforcement lines with Southern Region allies or the unfettered use of the Mediterranean. The NATO navies must now deal with a sea threat before turning their attention ashore. In essence, emerging Soviet sea power, when combined with the USSR's formidable ground- and land-based air strength, dramatically widens its horizons and permits the Kremlin to reach out for the many rewards that would accompany domination of the Mediterranean Basin.

NATO MARITIME POWER

Unfortunately, the West, either through overconfidence or apathy, has not seen fit to maintain its decisive edge at sea. In 1966 France withdrew from the military structure of NATO, questioning the role its naval forces would play. Certainly the small but impressive French navy would buttress the Southern Command's sea control effort, particularly with its two fixed-wing carriers, missile-firing surface ships, and attack submarines. In 1972 the United Kingdom abdicated its historical Mediterranean mission, and today there is not a single Royal Navy unit permanently deployed between Gibraltar and Suez. Moreover, the prospects for British reinforcements are not encouraging. Italy has added a significant number of modern ships to its fleet and puts an excellent ASW and AAW capability to sea, as well as a number of badly needed minesweepers. These units, combined with a small submarine, patrol aircraft, and helicopters make a welcome but nevertheless modest contribution to NATO's naval effort. Greece and Turkey, for economic reasons, have slowly reduced the size of their fleets, and every year fall further behind in the race for quality. They still have a predominant number of World War II vintage units in their inventories. If the issue is joined, they will play a gallant and crucial role in the Black Sea, Aegean, and entire Mediterranean but will be greatly outnumbered and outgunned by the Pact forces. In sum, the Soviet building program has greatly outstripped the indigenous navies in numbers, sophistication, and drive, which has predictably shifted the bulk of the naval burden to the U.S. Sixth Fleet.

While the U.S. Navy is undergoing a vigorous and continuing modernization program, its size has been halved in the last decade, and this drawdown has been reflected in its Mediterranean forces. Simultaneously, its expanding global responsibilities necessitated the withdrawal of a carrier battle group from the Mediterranean in 1980 to redeploy to the Indian Ocean. Numerically, the U.S. naval presence is the smallest it has been since World War II.

These developments notwithstanding, the American Sixth Fleet still represents considerable power. On an average day one carrier, its supporting surface units, a number of nuclear submarines, and an amphibious group with over 1,800 marines embarked will be deployed in forward areas of the Mediterranean. These units will be supported by several squadrons of maritime patrol aircraft and the essential afloat support. In a direct match-up with the Soviet *Eskadra*, the United States enjoys the advantage of seaborne air power, both offensive and defensive; an advanced underway replenishment capability; a deep-rooted seafaring tradition; superior training and technical expertise; and a wider command initiative for its commanders than do their Soviet counterparts. On the other hand, the confines of the Mediterranean ensure that U.S. units will not be easy to conceal and always will be in range of Soviet land-based naval aircraft and vulnerable to Soviet submarine attack. Thus, the numbers each side can bring to bear at the outset, the manner in which the first salvo is initiated, the extent of NATO participation, the ability to seal off choke points rapidly, and the availability of reinforcement all become extremely important in the tactical calculus.

While it is likely that today the alliance (assuming full participation) would ultimately prevail at sea, a genuine showdown in the Mediterranean between NATO and the Warsaw Pact would be hard-fought with considerable attrition and few guarantees of success. Whether U.S. naval air power would make a substantial contribution to the land campaign would depend first on the struggle for sea control. Even if the Soviets were ultimately contained or defeated, they might exact a high price. For example, if in the process they were to prevent the carriers from assisting the land battle and delay, or prevent seaborne resupply and reinforcement from reaching beleaguered ground forces, they would have dealt NATO strategy a severe, if not mortal, blow. Certainly, if the sea battle is lost, the fate of NATO's land forces on the southern front would be sealed. Instead of the marked superiority of the 1960s, the margin today is perilously narrow, and if this trend continues, it could be reversed in favor of the Kremlin. Just as Moscow's policies have produced an emerging and disturbing maritime challenge, NATO's failure to match the investment has enhanced the temptations for the Kremlin to flex its newly developed naval muscles in the Mediterranean.

THE OUT-OF-AREA MEDITERRANEAN

Accompanying the above events, severely complicating the strategic picture, and simultaneously lowering the flash-point temperatures have been the sweeping political changes of the last thirty years around the eastern and southern shores of the Mediterranean. Apart from the southern flank nations, there are fourteen countries on the littoral with 215 million inhabitants. The colonial system has disappeared altogether and a wide spectrum of governments and philosophies has emerged, from authoritarian regimes to genuinely representative governments, from assumedly Marxist to staunchly traditionalist, from pro-West to Soviet-supported.

By Western European standards many of the non-NATO countries are economically very poor indeed. The per capita income in several of the nations is well below 1,000 U.S. dollars per year. Literacy rates are generally low. Few have democratic and stable governments, and most have populations with at least 40 percent below fifteen years old. In Libya, this figure is a staggering 51 percent. By contrast, the under fifteen age group in France and Italy is only 23 percent. Clearly, feeding and finding employment for these burgeoning populations will pose serious future problems for their governments.

Relationships between many of these countries are often severely strained, if not openly hostile. Moreover, some are acquiring impressive arms arsenals that can turn friction into serious conflict. In short, grave political, population, economic, and social pressures will continue to test these fragile regimes. An inventory of the instability and potential crisis points around the shores of the Mediterranean leads to the conclusion that threats to peace of the southern front, perhaps to all of NATO, are not restricted to Europe but also lies south and east of the traditional NATO area of interest.

In turn, sea power plays a direct and instrumental role in this drama. There is considerable evidence that in the halcyon days of the Sixth Fleet its presence and influence were an important factor in maintaining stability in the region, particularly when local disputes threatened to escalate into violence. The emergence of Soviet sea power has altered the equation somewhat. Soviet units are now at the scene of every crisis, not only to evidence Moscow's political interest but also often to encourage rather than discourage friction. This fact highlights an important point: the ultimate alignment of countries on the littoral in periods of tension or war may well be influenced by the ability to control the sea. Just as with Western Europe, emerging nations today are tied to the Mediterranean, a reality they cannot ignore. It is sheer folly for NATO to expect the active and voluntary cooperation of these countries if the West cannot guarantee their livelihood and provide outside support.

In the event of a major conflict, such a vulnerability could be crucial. Not only does the West depend upon the resources of non-NATO countries, but also many of these nations now have sizable military capabilities that could play a significant role in the overall strategic equation. Syria, Iraq,

Israel, Egypt, Libya, and Algeria all possess impressive air power orders of battle, including numbers of modern attack aircraft that could strike sections of southern Europe as well as ships at sea. Their ground forces, although limited in size, cannot be ignored and are in some cases equipped with modern armor, artillery, air defense weapons, and personnel carriers. Perhaps more disturbing would be the accessibility of their facilities to outside powers. Libya has built a great many more air facilities than it can use. They are modern, well defended, and, if available to a sophisticated external force, could pose a disturbing challenge to the central and western Mediterranean. Such an offer to the Warsaw Pact would dramatically alter NATO's security calculations. The political changes that have taken place in the Near East and North Africa over the last thirty years have understandably raised the military ante in the Mediterranean.

Thus the ingredients are now present to make the Mediterranean an awesome battle zone once again. Such a struggle would be just as crucial, if not more so, to the onward path of Western civilization, as were so many past Mediterranean campaigns. It also would be more costly in both blood and treasure than any those waters have previously known.

CONCLUSIONS

If such a conflict is to be avoided, the NATO nations must acknowledge the dimensions of this new danger and redouble their combined efforts to preserve vital Western interests throughout the Mediterranean basin. Even with the enlarged fleet that the United States now plans to build, that strong but much-burdened country will not be up to the task by itself. During the 1980s and 1990s the major NATO powers must again devote more attention and substantial assets to their Mediterranean flank. In particular, those member nations fronting on the sea, including France, must be encouraged to exercise a more assertive and projective role in fostering Mediterranean-wide security. Greater maritime muscle also will enhance their capacity to exercise peacetime leadership among the volatile nations on the eastern and southern littorals, whose territory, weaponry, and energy resources have become new factors in the large power equation. The impending entry of Spain as a full partner in NATO promises a significant contribution to overall military capability, coupled with the advantages of the geographical position of that country.

At the same time, the central and northern allies, recognizing their stake in the area, must reconsider the sparsity of their current naval contributions. It is a time for realism and imagination as well as the development of innovative policies, perhaps including, for example, rotating peacetime naval deployments to the Mediterranean, enhancing security assistance to the nations of the southern flank, and designating naval and air reinforcements in crises.

As for the United States, the critical question is whether its global commitments and priorities will permit it to muster the necessary assets in the Mediterranean at the proper time. In sorting out its priorities, Washington must squarely face the Mediterranean's new economic and strategic significance. Only collective Western determination and strength can forestall conflict throughout the Mediterranean or, in the alternative, win the ensuing contest. It is imperative that the NATO allies meet the new and important challenge emerging in this ancient sea.

The Soviet Union in the Mediterranean

THOMAS H. ETZOLD*

In the Mediterranean, Soviet policy, posture, and activity over the past generation take their significance from a single fact of political geography: nowhere else in the world do NATO's traditional alliance concerns mesh so extensively, unavoidably, and decisively with "out-of-area" problems. The Mediterranean area may very likely become the most troubled in the world in the next two decades. Here, East-West competition in Europe, regional strife in the Balkans, Arab-Israeli conflict, the contest between political Islam and Arabism, and global U.S.-Soviet rivalry as well as the North-South struggle all intersect.

This intersection of issues now requires NATO to consider whether it can remain a regional alliance, or whether its members must engage in a global contest over developments well beyond Europe. Whether or not NATO as an alliance faces up to the foregoing question, the intersection of issues in the Mediterranean poses immediate economic, political, and military threats to the NATO states even in their local setting. As has often been remarked, NATO's European members, for the most part, depend on Middle Eastern oil but lack the ability to ensure its flow or significantly to influence its price (Great Britain and Norway are the exceptions). Politically, the Middle East and Southwest Asia continually threaten to become the graveyard of détente. Militarily, U.S. concerns with Southwest Asia, given form in the creation of the Rapid Deployment Force (RDF)—now the forces of the U.S. Central Command—could compel a reallocation of American military resources on a scale unprecedented since the first deployment of American forces to Europe in the early 1950s. As if it were a political and strategic black hole, Southwest Asia could absorb huge proportions of U.S. forces. Perhaps more important for NATO, planning for and employing the RDF in its Southwest Asia missions could decouple the United States from Europe's security more surely and more extensively than has been contemplated in the theoretical, political, and nuclear debates of the past twenty years.

The role of the Soviet Union in bringing about the situation and trends so far mentioned has been relatively modest. To a considerable extent, the familiar alarms over Soviet activities in the Mediterranean have constituted special pleading. It seems impossible within NATO to hold a conference on one country or one region without falling into a parochialism that focuses the

*The views expressed here are those of the author and do not necessarily reflect those of any agency of the U.S. government.

Soviet threat on the area under discussion. On rare occasions this may become the recognition of special opportunity; on most, it is no more than the assertion of special interests. With regrettable regularity, arguments and evidence concerning qualitative and quantitative improvements in the order of battle facing NATO's southern members are adduced to substantiate the need for an upgrading of the southern sector of NATO as against the center. Similarly, invocation of the Soviet threat to Southwest Asia justifies priority U.S. and alliance-wide investment in Southern Region defenses and forces on the dubious premise that it is necessary to discourage a diversionary attack on the Southern Region as an offshoot of a Soviet southward move.

Certainly, the record of Soviet claims in and around the Mediterranean has established that there is reason for concern. Even without the adornment of imputed motives and menacing master plans, Soviet attempts since World War II to construct positions of dominance throughout the Mediterranean area demonstrate that Soviet interests have been, and will probably remain, highly incompatible with those of the Western allies. The Soviets have endeavored to seize northern Iran; support insurrection in Greece; take away control of the Bosphorus and the Dardanelles from Turkey; lay claim to Turkish Thrace, the western shore of the Bosphorus, the Dodecanese Islands, Eritrea, Libya, and Malta; and make clients of Yugoslavia, Yemen, Syria, Iraq, Egypt, Libya, Somalia, and Ethiopia.

The record of Soviet activity in the Mediterranean appears often in the literature.[1] Its meaning, however, remains conjectural and therefore in need of more discussion. Hence, two related questions will be explored here: What does the Mediterranean signify in terms of Soviet concerns and goals, insofar as these can be fathomed? What have been the larger effects of Soviet military posture and strategy in and around the Mediterranean?

SOVIET CONCERNS AND GOALS IN THE MEDITERRANEAN

Soviet concerns and goals in the Mediterranean appear to be 1) primarily defensive in their political-strategic character, and 2) for the most part tangential to NATO. Despite the prominence accorded Southwest Asia, the

---

[1]See in particular Mohammed Ayoob, "Autonomy and Intervention: Super Powers and the Third World," in *New Directions in Strategic Thinking*, ed. Robert O'Neill and D. M. Horner (Boston: George Allen & Unwin, 1981); Robin Edmonds, *Soviet Foreign Policy 1962–1973: The Paradox of Super Power* (London: Oxford University Press, 1975); Benson Lee Grayson, *Soviet Intentions and American Options in the Middle East* (Washington, DC: National Defense University Press, 1982); Paul B. Henze, *Goal: Destabilization—Soviet Agitational Propaganda, Instability and Terrorism in NATO South* (Marina del Rey, CA: European American Institute, 1981); Bruce R. Kuniholm, *The Origins of the Cold War in the Near East* (Princeton: Princeton University Press, 1980); Richard C. Thornton, *US-Soviet Strategic Balance in the Middle East, 1977–1981* (Washington, DC: Institute for Sino-Soviet Studies, 1981); Adam B.Ulam, *Expansion and Coexistence: Soviet Foreign Policy 1917–73*, 2d ed. (New York: Praeger, 1974); Lawrence L.

Middle East, and the eastern Mediterranean in the last several years, this area remains at best secondary in importance, with Eastern and Western Europe continuing as the principal areas of Soviet strategic concern.

Soviet activities in the Mediterranean area since the Second World War fall into three distinct periods. For a decade after the war, the Soviet Union simply exploited postwar turmoil and regional strife where possible: Iran, Greece, Turkey, Yugoslavia. With the controversy over support for the Aswan High Dam and the Suez crisis in 1955–56, a more pointed and concerted Soviet policy took shape as the Soviets commenced a sustained effort to build a permanent position of influence among Arab states. Equally important, the Suez crisis—and especially the way the allies handled it—forged long-term links among NATO issues, Middle East conflict, and the West's relations with Arab states. Between 1970 and 1974 this phase of Soviet Mediterranean involvement gradually gave way to still another. As Egyptian-Soviet relations began to fluctuate, Soviet interest and activity intensified in the so-called "outer ring" states of Somalia, Libya, Yemen, Aden, Ethiopia, and, to a lesser extent, Iran. In 1971, almost as if anticipating the rupture of relations with Egypt in 1972, the Soviets prepared a fall-back position in Somalia.

The 1973 October war exerted a catalytic effect on the area's issues. It stimulated rapid development of Islamic political radicalism through the remainder of the 1970s, excited revival of intense Arab nationalism, and transformed both the world energy market and the political and financial significance of the Gulf states. Moreover, it triggered U.S.-Soviet military posturing and counterposturing to a degree of menace not seen since the Cuban missile crisis of 1962. But the October 1973 war may have been less notable for what it brought about than for what it did not. The anti-U.S. coalescing of Arab states and the use of oil as a weapon failed to reverse the worsening of Soviet-Egyptian relations. The petroleum exporting states of the Persian Gulf and elsewhere certainly aggravated the economic and political problems of Western industrialized states, as well as those of Third World nations, but remained for the most part independent of Soviet policy and influence. As in its earliest days, the Soviet Union could find friends only among political outcasts on the radical fringes of the Arab world—the militarily inept, politically isolated Syrians, and the erratic, blustering Libyans. Soviet efforts to develop longer term positions of advantage in Iraq, South Yemen, Somalia, Algeria, and Ethiopia fell far short of expectations, with only that in Ethiopia offering much substance by 1983. There the Soviet Union gradually extended its ability to operate naval forces in the Indian Ocean.

In order of increasing relevance to NATO itself, here follow what the record suggests are the main Soviet goals in the Mediterranean area:

Whetten, *The Soviet Presence in the Eastern Mediterranean* (New York: National Strategy Information Center, 1971); and James T. Westwood, "The Soviet Union and the Southern Sea Route," *Naval War College Review*, 35 (January–February 1982): 54–64.

1) Promote the Dynamic Status Quo

The Soviets consider the status quo to be dynamic, embodying trends and movements ultimately favorable to the emergence of socialism and people's democracies throughout the world. Continuation of rapid decolonialization, "wars of national liberation," class struggle, and civil strife between the Right and Left elements of the political spectrum all are legitimate, ordinary ingredients of "normal" international relations at the present stage of the world sociopolitical development. The Soviet Union employs its political, economic, and military resources to sustain these activities and to make it difficult for the Western powers to impose an "unnatural" stasis. What the West calls stability, the Soviets consider reactionary repression—neither normal nor a status quo but the suppression of the natural components and dynamics of a genuine, progressive status quo.

2) Limit U.S. Influence in the Arab World

The Soviet Union, like the United States, desires to limit the influence of principal political adversaries to the extent possible, whenever possible. The location, resources, and religious culture of the Arab states give particular reasons for Soviet concern. In the twentieth century no state located to the south of the Soviet Union has been capable of defeating the Soviet Union in war. China has been the only significant military threat from that direction. For the Soviet Union, the postwar reduction of British and French power in the Mediterranean, Middle East, and North Africa offered an opportunity to rid the area of the outside great-power presence that posed the only possible strategic threat. By taking up the roles relinquished by the British and French, the United States could only damage Soviet interests. The growing long-term importance of Middle Eastern oil also has played a part in Soviet calculations and policies toward this area, although exactly what part remains a matter of debate. Furthermore, the practical political interplay between Arabism and Islam has gained importance. With its 37 million Turkic peoples and 6 million more Moslem religionists in the central Asian republics, the Soviet Union has been especially sensitive to great-power involvement in Arab politics.

3) Limit the Value or Usefulness of the U.S. Tie with Israel

A companion to the foregoing objective, this point holds special interest in its own terms as well. Apart from continuing attempts to find advantage for the Soviet Union and injury for the United States through manipulation of Arab-Israeli conflict, the Soviet Union has come to fear the military and strategic implications of U.S.-Israeli cooperation. In a number of easily imaginable situations, Israeli bases and facilities could prove both more available and more useful than those located on the territory of NATO members. Under many conditions, Israeli military support, perhaps even including employment of Israeli forces in conjunction with those of the United States, would be more certain than would any military cooperation from among NATO's Mediterranean states.

In the Middle East, in any Southwest Asia contingency, or in the eastern Mediterranean, U.S.-Israeli military ties weigh heavily in the Soviet political, strategic, and tactical calculus. Politically, American ability to influence Israeli policy, however unsatisfactory in extent from an American point of view and however sensitive from an Israeli viewpoint, nevertheless has formed the basis of U.S. influence in the Arab world for the past fifteen years. Furthermore, the Soviet leadership considers Israel the only state in the world whose interests come close enough to those of the United States to support a genuine and effective political-military, strategic partnership. Not even the interests of the NATO states appear to the Soviets to coincide with those of the United States closely enough to permit a true durable partnership. Ideologically, the Soviets insist that no nation's interests really parallel those of the United States, but they refer to U.S.-Israeli cooperation as "partnership in action."[2] The U.S.-Israeli connection has assumed a political, strategic, and ideological significance for the Soviets, in many respects far greater than that accorded the relationship even in the bitterest U.S.-NATO discussions over Middle East policy.

4) Retain Access to Middle Eastern Oil

Soviet objectives in regard to oil have received considerable discussion in the last ten years and have become the subject of modest contention. The foregoing formulation represents a conservative statement of the long-term Soviet goal, but it does not exclude occasional temporary expansions of hope or tactical adoption of more offensive-sounding positions. Several points deserve special note in this matter. First, it continues to be widely known that in about 1977 the Central Intelligence Agency concluded that the Soviet Union would become a net oil importer in the 1980s; it is much less widely known that it later reversed this judgment. At present, the Soviet Union seems unlikely to need Middle East oil for its own uses, although East European economies might. How much of Eastern Europe's energy requirement the Soviet Union will be able to meet from internal production and how much Eastern Europe, as well as Western Europe, will depend on Middle East oil in coming years cannot now be determined.

Second, Soviet writings indicate no clear intention to deny the West access to Middle East oil. Recent Soviet comment stresses the need to maintain everyone's access to it, including that of the Soviet Union and Eastern Europe. Soviet writers now argue that it is "impractical and counterproductive" to cut

---

[2]John Lenczowski, *Soviet Perceptions of U.S. Foreign Policy* (Ithaca: Cornell University Press, 1982), p. 208.

off the West from raw material resources, in part out of recognition that Soviet support of the Arab oil embargo had strengthened antidétente feeling in the United States. At present the emphasis is on "worldwide cooperation, especially between the United States and the Soviet Union . . . to avoid any worldwide energy shortage."[3]

Third, many knowledgeable people nevertheless are willing to conclude that Soviet objectives are more far-reaching and inimical in respect to Middle East oil. The Commission on U.S.-Soviet Relations, sponsored by the Council on Foreign Relations, recently has argued that in Southwest Asia the Soviets aim "to establish influence for its own sake; to prevent a renewed Western military presence; to acquire potential capacity to deny, constrict, or control Western access to the area's energy; and to assure Soviet access if or when it comes to have major energy needs."[4] Interestingly, current, although private, Chinese evaluations of Soviet strategy and intentions also stress Soviet desire to control the Middle East and relevant sea lines of communications. Yet Geoffrey Jukes has suggested some limits to grand geopolitical reasoning about Soviet intentions in the area:

> It has been suggested that [Soviet action in Afghanistan] is a part of a plan to encircle the Gulf, and that the next step will be an invasion of Pakistan, especially Pakistani Baluchistan, with the aim of seizing a base for naval forces. . . . This would seem rather an unlikely form for Soviet influence-seeking to take. On one hand, it would do little or nothing to threaten oil traffic that could not be done by air action from airfields in Soviet Transcaucasus or the airfield at Shindand in Western Afghanistan. On the other hand, it would undoubtedly open the doors of many Indian Ocean countries to the establishment of a strong Western counterpresence which they are at present reluctant to contemplate.[5]

Finally, swings in Middle East and Persian Gulf politics or events may well embolden the Soviets from time to time, giving them reason to hope, or even to expect, that Western access to Persian Gulf oil will be reduced or cut off. In the long run though Soviet policy and interests seem to turn on 1) maintaining a regional and international regime that keeps Middle East oil available to all potential purchasers, and 2) assuring that the price of oil stays relatively high. As one of the world's largest oil producers and exporters, the Soviet Union is well served by high energy prices. The collapse of OPEC's price and supply cartel early in 1983 produced considerable Soviet anxiety about the prospect of a rapid fall in prices and a corresponding reassertion of

Soviet oil policy

---

[3]Roger Hamburg, "Low-Intensity Conflict: The Soviet Response," in *U.S. Policy and Low-Intensity Conflict: Potentials for Military Struggles in the 1980s*, ed. Sam C. Sarkesian and William L. Scully (New Brunswick, NJ: Transaction Books, 1981), pp. 165, 166, 167.

[4]The Commission on U.S.-Soviet Relations, *The Soviet Challenge: A Policy Framework for the 1980s* (New York: Council on Foreign Relations, 1981), pp. 19, 20.

[5]Geoffrey Jukes, "Soviet Strategy, 1965–1990," in *New Directions in Strategic Thinking*, ed. Robert O'Neill and D. M. Horner (London: Allen & Unwin, 1981), p. 62.

neocolonial control of Western banks and corporations over the resources of the oil-producing states. Interestingly, Soviet oil prices fell to $27 per barrel in March 1983, one day after OPEC finally agreed on a base price of $29 per barrel for Saudi Arabian light crude.

5) Deny to the United States any Support from Its NATO Allies in Coping with Problems in the Middle East, Southwest Asia, the Persian Gulf, and North Africa

The Soviet Union has long regarded the United States as a calculating, deliberate troublemaker. Soviet foreign policy specialists accuse the United States of pursuing a policy of "controlled tension," of using the "constant threat of war" to keep superpower and East-West relations continually on the verge of crisis or war. The principal instrument in this effort, so the Soviets aver, is the American military, dangerously and incitingly deployed globally in a "military base strategy." The United States has suffered from the incurable illness of "base mania," acquiring approximately two thousand bases in thirty countries. (The Soviets apparently count everything as a base, down to and including a radar station with a handful of men.) These bases are used to "pursue a policy of neo-colonialism," "pressure junior partners in military blocs," "support unpopular regimes," and conduct war. Typically, or so Soviet writers claim, Americans employ the constant threat of war through "gunboat diplomacy" and military maneuvers.[6] Soviet policy and strategy worldwide aim to render unavailable the bases and facilities from which U.S. forces must operate in forward areas. For obvious reasons, this is particularly important in locations such as the eastern Mediterranean. Amid the Mediterranean's intersecting problems, the Soviets benefit from and attempt to exploit natural divisions of perspective, interest, and policy among the allies. In turn, the advantage that accrues to the Soviets when, for example, Turkey declines to receive American aircraft en route to rescue hostages in Tehran, intensifies alliance debates as to whether immediate or ultimate considerations should dominate response in crisis.

6) Make NATO Politically and Militarily Incapable of Concerted Action in Crisis and War

In Soviet ideas on how World War III might come about, the Western allies invariably force war on the Soviet Union and its allies. This is more than self-justification or self-delusion. Beyond doubt, the Soviet leadership fears both the political intentions and the military capabilities of the United States and its allies. The pretext upon which the West seeks war, the extent to which the United States compels its allies to participate, and the exact

[6]See Lenczowski, *Soviet Perceptions*, pp. 155ff.

circumstances and unfolding of crisis and beginnings of war vary. Further, as many specialists have observed, the Soviet concept of "deterrence" is not at all equivalent to that held by the West. Nevertheless, they seek to avoid war and to make it less likely, in part through the readiness of Soviet military forces to win if engaged, and in part through political efforts to deprive NATO of consensus on perceptions of Soviet threats, to emphasize differences in the interests of the United States and those of all other NATO members, and to sow doubt about American military power and political reliability. Moreover, Soviet analysts assert that the United States has recognized that it "assumed a burden of military pledges which was beyond its power" and therefore has sought out of necessity, "while maintaining a prominent role in NATO and other imperialist blocs to shift a considerable share of the burden and responsibility of preparing and waging certain types of wars onto its allies."[7]

In the early 1980s, after more than a generation of postwar effort in the Mediterranean, the Soviet Union could count few assets and successes. Except for its Balkan satellites, the Soviet Union owned no militarily significant and politically reliable allies in the Mediterranean or its neighboring regions. In peace unimportant and in war poorly located, Syria could scarcely qualify as an important partner, although Soviet bases that might be installed on its territory could threaten the Sixth Fleet. Like Castro's Cuba, Libya's Qaddafi possessed more arrogance than power. An annoyance, Qaddafi, through his behavior, made Libya a possible target, not a regional influence, in crisis or war. Early in 1983 the Soviet leadership once again prepared to swing military and political support away from Iran to Iraq in those states' continuing struggle. However expedient and opportune, this shift underscored the reality of failure rather than the hope of gain.

In the Soviet outlook, disappointment at the lack of substantial allies joined with distress over the threat posed by renewal of American political and military will. As Roger Hamburg has noted,

> the wish to avoid pushing the United States or its NATO allies into augmented military buildups or diplomacy adverse to Soviet interests has been a constraint on Soviet political-military behavior. The recurring Soviet insistence that their military augmentation is no threat (as they put it, the 'so-called Soviet military threat') is not only an attempt to allay Western suspicions when the Soviets take a particular action or build up forces; it also reflects real concern over an exaggerated Western response to what in Soviet eyes may be a mere tactical move or maneuver in a never-ending struggle.[8]

Events in Afghanistan and Iran in 1979 and 1980, together with the election year's coverage of defense issues, revived American political and military will to a degree not seen since the mid-1960s. Détente withered, and

---

[7]Ibid., p. 173.

[8]Hamburg, "Soviet Response," p. 156.

with it Soviet hopes of obtaining American technology on terms that would ease the perennial crisis of the Soviet war-mobilization economy, as well as Soviet détente-based expectations of increasing economic advantage in Western Europe and a declining U.S. presence and involvement in various Third World areas. Further, however grave the shortcomings of RDF ideas and structures, when announced the RDF seemed an extraordinary threat to the Soviet Union's interests outside Europe. Taking the United States at its word, Soviet military writers emphasized the creation of "special strike forces" for intervention. According to Soviet analysts, these consisted of all American forces not assigned to NATO: 600,000 ground, naval, and air personnel including the entire Marine Corps, two army divisions, airlift, thirty navy aircraft carriers, and ancillary forces.[9]

Moreover, OPEC's inability to maintain high prices and constricted supply in the oil market suggested to the Soviets that the United States and its Western associates might succeed in reasserting neocolonial control over the Gulf's resources, perhaps recouping much of the power and wealth lost in the 1970s. NATO's own intra-alliance problems may have offered the Soviets some cheer in an otherwise cheerless moment, but they could take only comfort, no credit, for NATO discord and uncertainty in addressing the intersecting Mediterranean, Balkans, Middle East, Persian Gulf, and Southwest Asian problems.

SOVIET STRATEGY AND FORCES IN THE MEDITERRANEAN

In the preceding discussion of Soviet goals and concerns in and around the Mediterranean, military dimensions received only passing attention. Both intellectually and politically it is unduly inappropriate to segregate political from military matters when dealing with large strategic concerns. Yet, in something of a paradox, proper understanding of military developments bearing on the Mediterranean strategic situation depends on careful distinction between the uses of military forces in peace and crisis and, in contrast, their use in war. The issue is simple, though perhaps not really straightforward because the always complicated relationship between military actions and political results differs fundamentally in war from what it is in anything short of war. According to venerable strategic theory and precept, all war is political; military forces are always employed in pursuit of politically defined aims. In military operations short of war, however, political outcomes are the immediate goal and direct result. War requires the pursuit first of military results,

---

[9]Frank N. Trager and William L. Scully, "Low-Intensity Conflict: The U.S. Response," *U.S. Policy and Low-Intensity Conflict*, pp. 187–88. The number thirty referring to aircraft carriers is either a misprint or a mistranslation of an article in *Red Star* by Colonel M. Ponomarev ("Strategy of the World Policeman," 12 August 1979). The aircraft number specified would be about right for twelve or thirteen carriers.

whatever the ultimate political consequences envisioned. All of this would remain abstract and perhaps irrelevant if it were not possible to produce substantial political effects in peace and crisis with forces that would prove militarily ineffective in war. The importance of Soviet military development in the Mediterranean over the past generation resides in this single point: the Soviet Union has learned to exert political influence in peace and crisis with forces that would prove militarily inferior in war.

Because Soviet military posture and activity in the Mediterranean possess more peacetime political importance than wartime military potential, the Mediterranean military situation since World War II has exhibited more fundamental continuity than change. Such a judgment, though, depends on the perspective and reference points chosen. Precisely because the relationship between military power and political results is so complex, it is extremely difficult to predict, measure, and assess what takes place. Recent emphasis on what has changed grows out of a preference for measures of the Soviet order of battle, deployed ship days in the Mediterranean, or crisis surges by the Fifth *Eskadra*. In such respects, compared to, say, 1960, circumstances seem to have changed a great deal and all for the worse, although from 1970 to the early 1980s Soviet Mediterranean deployments have averaged about the same, with peaks in crisis. In a long perspective both the degree and adversity of change seem diminished.

The Soviet Union's present role in the Mediterranean differs little from what it has been since the end of the war of Napoleon. The Soviet Union, like Russia before it, is one of the most important European—and therefore Mediterranean—powers, arguably now the most significant. Yet its access to the Mediterranean remains restricted, its strategic interests and posture are primarily defensive, and its forces are unable to secure wartime use of the Mediterranean or to deny use of the sea to likely enemies. Moreover, the Soviet Union lacks important or useful allies in and around the Mediterranean.

The foregoing observations are preliminaries to a description and discussion of Soviet military enterprises in the Mediterranean. They are meant simply to establish that these activities, while serious, are something less than tragic.

Over the past generation the Soviet Union has pursued a variety of military activities in the Mediterranean, primarily for their peacetime political effects rather than for their wartime promise. It has deployed more numerous, more capable ships more of the time and in more of the Mediterranean than in earlier years. It has patched together and, with mixed results, attempted to maintain an overseas infrastructure of bases, facilities, anchorages, and ship repair sites around the Mediterranean littoral, on some islands, and on the Red Sea. Since 1977 Algeria, Libya, Syria, Iraq, and Ethiopia each has received more than $1 billion in armaments from the Soviet Union. Soviet military personnel serve as advisers and sometimes more in each of these countries as well as in South Yemen. Seeking still closer ties, the Soviet Union has signed treaties of friendship and cooperation with Syria, Iraq,

South Yemen, and Ethiopia, although these fall somewhat short of military alliances.[10] (The only Soviet military alliance outside Eastern Europe is with North Korea.)

Since the early 1970s Soviet forces have taken more substantial, more prominent, and more confrontational roles in the region's crises. Soviet pilots flew against Israeli air crews in 1970 and 1971, though at a time when the United States was attempting to restrain Israeli air operations against Egypt. Indeed, Soviet willingness to use "force in its long-term quest to outflank foes and promote its vision of the world" has impressed even the most sophisticated foreign affairs experts as a signal development in the past fifteen years.[11] In the October war the Soviet airlift to support Egypt commenced before the American airlift to resupply Israel, in spite of the tenuous course of Soviet-Egyptian relations over preceding months. In 1975, reacting to the crisis in Angola, the Soviets sent a naval surface action group to show the flag off the African coast, apparently in the mistaken belief that the aircraft carrier *Saratoga* had been dispatched from the United States for the same purpose. Similarly, early in 1983 Soviet forces accompanied the SA-5 antiaircraft missile batteries deployed to Syria in the wake of Israel's success against SA-6 batteries there. Indications that Soviet forces were engaging in more than advisory activity in Libya also continued to appear from time to time, although specific details and evidence remained difficult to obtain.

Throughout the 1970s routine Soviet naval presence, while not very threatening, had pervasive political and psychological effects. It contributed substantially to efforts to compel all European states to accord it what one author referred to as a "*droit de regard*."[12] Soviet naval forces did not establish the ability to decide issues directly and conclusively wherever and whenever they appeared. But they made it increasingly difficult to apply Western power without first considering Soviet interests. Offsetting naval presence did not necessarily equal offsetting political influence, although it did alter the political context and thus the stakes of situations in which both the Soviets and the West evinced an interest.

Increasing naval deployments, high and continuing arms transfers, and a somewhat larger profile in regional conflicts represented no fundamental change, either in Soviet desire for war or in its ability to conduct war in the Mediterranean area if one should come. Soviet leaders in the early 1980s retained a sense of extraordinary insecurity as they considered their country's strategic situation. The strategic linkage of southern Europe to the Middle East, the Mediterranean, the Persian Gulf, and Southwest Asia carried a renewed threat of encirclement by the West from the Bosphorus to Vladivostok. China's role as the "17th member of NATO," together with its close

---

[10]*Soviet Military Power 1983* (Washington, DC: Government Printing Office, 1983), pp. 84, 85.

[11]Commission on U.S.-Soviet Relations, *Soviet Challenge*, p. 8.

[12]Johan Jørgen Holst, "Deterrence and Stability in the NATO-Warsaw Pact Relationship," *New Directions in Strategic Thinking*, p. 92.

ties to Pakistan, made this fear all the more pronounced after the invasion of Afghanistan in 1979. As mentioned earlier, Soviet leaders also worried about the effects of Southwest Asian and Middle Eastern religion and politics among central Asian people within the USSR. Moreover, in the early 1980s increasing discussion of the possibility of launching NATO counteroffensives into Eastern Europe, in the event of major war with the Soviets, heightened concern over the Mediterranean's potential use to the West in war.

Soviet naval operations, arms sales, advisers, and crisis behavior, however, certainly engendered both apprehension and misapprehension among Western political-military analysts. These surfaced particularly in public discussions of the Mediterranean naval balance and its implications; in the possibility that NATO's Southern Region might be the object of a diversionary attack in the event of a Soviet drive into Iran; and in speculation about the possible attraction Soviet strategists might feel toward employing tactical nuclear weapons against states of NATO's Southern Region.[13]

In important dimensions the Mediterranean naval balance has been treated indifferently, though not with indifference, in the literature. Perhaps people have been too ready to accept appearances, and perhaps the United States and its allies have not done enough to promote a wider understanding of that balance. The problem calls to mind former Chief of Naval Operations Elmo Zumwalt's story about the visit of Secretary of State William P. Rogers to Generalissimo Franco early in the 1970s: "The conversation went something like this. ROGERS: 'Are you concerned about Soviet naval strength in the Mediterranean, Your Excellency?' . . . FRANCO: 'Si.' Pause. ROGERS: 'Would Your Excellency care to elaborate on the reasons for your concern?' . . . FRANCO: 'No.' End of interview."[14] But it was not the end of the problem.

It is unwarranted to assume that the Sixth Fleet has been neutralized by Soviet naval forces, and that it would go down to ignominious defeat early in war. Soviet anticarrier warfare missions and doctrine do not necessarily call for the sortie of large proportions of the Black Sea fleet into the Mediterranean. In addition, Soviet interest in controlling the Bosphorus, frequently misinterpreted in the West, has almost nothing to do with ensuring the ability of Soviet forces to transit into the Mediterranean, and almost everything to do with preventing American or NATO forces from offensive intrusion into the Black Sea. Moreover, it is incorrect to interpret Soviet doctrine for coordinated all-arms attacks on aircraft carriers at the outset of war as meaning the immediate or early employment of nuclear weapons at sea. Although the threat embodied in Soviet naval aviation is considerable, it is not as if Western

---

[13] Richard B. Remnek, *Constraints on U.S. Military Power in Southwest Asia* (Carlisle Barracks, PA: U.S. Army War College, 1982); *NATO's Southern Flank, the Mediterranean, and the Persian Gulf* (Marina Del Rey, CA: European American Institute, 1982), especially the papers by Paul Olsen and Rolf Friedemann Pauls.

[14] Elmo R. Zumwalt, Jr., *On Watch* (New York: Quadrangle/Times Books, 1976), p. 457.

military professionals were too inattentive or incompetent to have evolved technology and tactics of their own to counter an improved adversary force.[15]

Those willing to concede that the Sixth Fleet might at least survive, and perhaps give good account of itself against Soviet forces in war, nevertheless bemoan the Sixth Fleet's newfound inability to bring air power to bear in support of land campaigns.[16] To some extent, U.S. naval aviation now offers to deliver less and less relative weight of ordnance against land targets but not for the generally cited reason that a carrier's aircraft all must be employed in defending the carrier itself. The carrier's potential contribution to ground campaigns has been diminished in part by the expanding air order of battle in both the East and West. Just on the basis of numbers, an aircraft carrier's 24 to 42 air-to-ground capable aircraft (the exact number depends on the mix of aircraft making up the carrier air wing) could make a more dramatic contribution to a campaign in which each side had 600 aircraft available than one in which each side might have 1,000 or more aircraft to employ. The attack contribution of naval aviation from the late 1950s through the late 1960s was largely thought of in terms of delivering tactical or theater nuclear weapons. Removal of the aircraft carriers from the Single Integrated Operations Plan (the nuclear war targeting plan) in the late 1960s appeared to alter the carrier's potential influence on great ground campaigns on the Continent.

With the A-7 program in the late 1960s and early 1970s, naval aviation adopted light attack aircraft as a mainstay of the carrier air wing. The light attack program offered significant improvements in reliability and maintainability of aircraft and also had the attraction of joint adoption by the air force, in a somewhat modified version. But when weight of ordnance delivered or deliverable is the criterion, light attack is likely to be found wanting. This latter problem, unfortunately, constitutes a present and future concern as well as a historical note.

The effect of naval aviation on the larger war will probably not be greatest as measured at the Forward Edge of the Battle Area (FEBA) or Forward Line of Troops (FLOT); that, indeed, was probably always the wrong place and way to measure its contribution. It is by no means evident that action there is the only useful support that naval aviation can offer Greece and Turkey, or other NATO countries in other NATO regions. The problem of identifying the best ways in which to use naval aviation is enduring, as it is with all other forces because best use depends on changing technology, tactics, threats, and targets. At the moment, the potential contribution of naval

---

[15]On Soviet concerns in the eastern Mediterranean, Turkish Straits, and Black Sea, see especially Michael MccGwire, "Soviet Military Thought: Its Implications for Maritime Warfare," unpublished manuscript, June 1983.

[16]Such concerns are expressed in Whetten, *Soviet Presence in the Mediterranean*, p. 34; Kenneth Hunt, "The Development of Concepts for Conventional War in the 1970s and 1980s," *New Directions in Strategic Thinking*, p. 186; and Maurizio Cremasco, "Military Options for the Security of the Southern Flank," *NATO's Southern Flank*, p. I–2.

aviation to gaining theater air superiority probably outweighs its air-to-ground role. In addition to bombers, aircraft carriers also bring surveillance, antisubmarine warfare, battle management, and air superiority systems to modern war to supplement surface and subsurface forces of increasing lethality. The object is not simply to "support" engaged allies but to defeat opposing forces. Properly employed, naval aviation can perform significant and useful missions with a high degree of success. In conjunction with other maritime and air forces, naval aviation, in war, will make a substantial contribution to the rapid defeat of Soviet naval and air forces in and around the Mediterranean.

In still other aspects, current discussions of the Mediterranean naval balance and possible naval war manifest an unwarranted pessimism that leads to flawed assessments of problems and correspondingly inappropriate solutions or strategies. In war, Soviet Mediterranean naval forces, especially surface forces, are unlikely to retain significant combat capability beyond the first few days. The survivability of Soviet submarine forces will be a direct function of their employment; the more they engage, the shorter their expected life span. The Soviet Union will encounter great difficulty in augmenting or reinforcing these forces in wartime. Without the ability to rearm while deployed, Soviet undersea forces have a real but sharply limited capacity to sink military or commercial shipping in a Mediterranean war. It is not at all clear that NATO's interests would be best served by denying Warsaw Pact forces access to the Mediterranean as war approaches, as some NATO experts advocate. In the Mediterranean these forces are more accessible targets than they would be in the Black Sea or, in the case of submarines, in open ocean transit. In any case, Soviet doctrine for war against shipping has long since recognized that destroying terminals of commerce, resupply, and reinforcement can be more effective than sinking ships one by one if and as they are encountered on the sea. Hence, the problem of protecting sea lines of communication increasingly depends more on air defense of ports, harbors, and modern cargo handling equipment, as well as on effective onshore security forces, than on sweeping sea lanes or escorting convoys.

The problem of sea lines of communication in many ways taxes the Soviets as much as, if not more than, the West. Gradually, Western analysts have begun to realize the importance to the Soviets of their sea line of supply across the Black Sea from the Odessa area to Bulgaria. In war against the Southern Region, this route would be crucial for campaigns in Thrace and in the Balkans. Further, the Soviet "southern sea route" from Black Sea ports through Suez, the Indian Ocean, and around to Vladivostok in large part sustains the Soviet Far East. How this might matter in war remains a subject of disagreement. The point is that wartime use of the seas in the Mediterranean is no one-sided issue, offering only threats to the West. There are opportunities as well.

The connection between Southwest Asia and southern Europe also has led to some misapprehension. It has become commonplace to acknowledge the possibility that the Soviet Union might attack NATO's Southern Region to divert Western, and particularly American, forces from opposing a southward drive. This point so misrepresents Soviet strategic views that it should not go undiscussed. In one sense, it represents a commutative European rendering of the American strategic nightmare: concurrent wars in Europe and in Southwest Asia, with large forces maldeployed in Southwest Asia when the Soviets reveal their "main aim" in Europe. But strategic logic is not necessarily commutative. Conceivably, Southwest Asia could be the target of a feint or diversion in a war on Europe. The idea that Europe might be a sideshow, a diversion from a Soviet axis of strategic advance in Southwest Asia, however, simply does not square with Soviet perspective on the prospect of war with the West and on the relative importance of Europe and Southwest Asia.

In the Soviet view of any major war, Europe remains the geographic area of primary concern. The opening of hostilities in Europe would supersede in importance any other action contemplated or in progress. It could not be a strategic ruse to ease movement toward goals elsewhere. Moreover, the Soviets do not now expect to attack some members of NATO without at least the United States actively in opposition. In turn, war with the United States, or with the West by definition, constitutes a conflict between opposed social systems—a global war, World War III, literally a war for the future. Socialist forces for progress must win that war when it comes and will undertake it only when forced to do so or when confident of victory but, in any case, not as a second- or third-order action. All of this leaves aside, for the moment, the larger question of whether the Soviets intend or expect to pursue whatever goals they have in Southwest Asia by resorting to war. Although the Soviet military has undoubtedly developed contingency plans for moving into northern Iran, Soviet leaders continue to regard war itself as unwelcome, an enterprise fraught with uncertainty, unpredictability, and uncontrollability.

Recent discussions of the strategic junction of Southern Region concerns with those of Southwest Asia have also reflected abiding misconceptions about Soviet views on the use of nuclear weapons, especially in theater warfare. In part, this reflects an inevitable time lag between evolution of doctrine, its discussion and confirmation in professional military literature in the Soviet Union, that literature's reading, translation, and, when necessary, eventual declassification or dissemination in the West. Yet open sources indicate that over the past decade Soviet thinking on the role and use of nuclear weapons has undergone a monumental shift. It is neither professionally nor politically adequate to say, as one SHAPE analyst recently wrote, that "it is unclear whether the Soviet Union would benefit militarily from escalating to nuclear weapons in a Southwest Asia or Southern Region operation."[17]

---

[17]Paul Olsen, "Some Considerations in Designing a Strategy for NATO's Southeastern Flank," in European-American Security Institute for Security Research, *Workshop on NATO's Southern Flank, the Mediterranean and the Persian Gulf*, vol. 1, *Working Papers* (Bagnoli, Naples, Italy: NATO Headquarters, 21–23 September 1981), p. V-10.

In contrast to Soviet professional military discussions of some years ago, over time Soviet attitudes toward the employment of nuclear weapons, even at theater level, have become profoundly cautious. Leonid Brezhnev's repeated assurance that the Soviet Union would not be first to employ nuclear weapons was more than mere propaganda. It reflected the sober conclusion that Soviet interests would be best served by attempting to attain military objectives conventionally. Special weapons of all descriptions and sizes remain under tight central control. Theoretically, Soviet military professionals continue to think of nuclear weapons as an element in a continuum of weapons and warfare. There is no "firebreak" between conventional and nuclear weapons in Soviet theory, so that in an abstract sense the Soviet military considers all wars nuclear, even though nuclear weapons may not yet have been employed.

This theoretical approach reflects not only an inclusive logic but also a political and military judgment that the West's conventional force inferiority, coupled with its anti-Soviet, anti-Socialist intentions, will result in Western attempts to employ nuclear weapons in war against the Soviet Union. During wartime, therefore, as Secretary of Defense Caspar Weinberger has written, "the main mission of Soviet intelligence is to determine the West's nuclear attack intentions." Soviet strategic defense missions, which hold highest wartime priority, place extraordinary weight on "destruction or neutralization of as many of the West's nuclear weapons as possible on the ground or at sea before they can be launched," this by means of non-nuclear weapons.[18]

Soviet concern over the potentially decisive effects of nuclear weapons employed on a large scale in the Armageddon-like "war between opposed social systems" bars the Soviet military from expecting to employ nuclear weapons at war's outset simply as another component of all-arms attacks. Release and use of nuclear weapons at sea or in secondary theaters—and compared with the central front, all else is a secondary theater—is extremely unlikely to precede a decision to employ nuclear weapons on the central front. The only plausible exception might be in the case of China: facing a war in the West, the Soviet Union might attempt to disarm or dissuade China in order to forestall Chinese entry into the war at an inopportune moment.

CONCLUSIONS

What, then, to conclude about the Soviet Union in the Mediterranean in the context of NATO concerns? First, whether in coming to terms or coming to grips with the Soviet role, posture, and prospects in the Mediterranean, neither particularism nor special pleading will serve NATO well. To the extent that Soviet actions and intentions in the Mediterranean threaten NATO, they threaten all members, at least so it is argued here. Politically, the Council on

[18] *Soviet Military Power 1983*, p. 14.

Foreign Relations probably forecast correctly that NATO members would respond to out-of-area concerns bilaterally or severally in coming years rather than with alliance-wide concurrence and coordination.[19] But NATO will pay a high price if the United States must deal unilaterally with perceived military threats in the Persian Gulf and Southwest Asia.

For the present, American investment in forces rapidly deployable to the Gulf must come at the expense of reinforcements now programmed for NATO. The most recent authoritative estimate suggests that deployment of U.S. forces to Southwest Asia could reduce the American contribution of ground forces to NATO by 20 to 33 percent, which would diminish NATO's overall balance of forces against the Warsaw Pact by 6 to 12 percent at about sixty days into a war. In coming years, as such figures suggest, expansion of the forces of the Central Command, as planned by the Reagan administration, will sharply limit American ability to provide modern equipment, military aid, and additional forces to European NATO members, both in peace and in war.[20]

Moreover, NATO's reluctance to grapple more directly with out-of-area problems has greatly strengthened the arguments of American advocates of a drastic reorientation of U.S. security policy. Atypical only in its elegance of formulation, Jeffrey Record's statement in mid-1982 captured a growing American mood:

> A revolution in America's strategic outlook and military posture is dictated by two seminal events . . . the steady and apparently irreversible disintegration of the North Atlantic Treaty Organization as an instrument capable of mustering an adequate collective forward defense of Western Europe . . . [and] the emergence of a host of new threats to vital United States security interests in Southwest Asia and in other areas outside Europe where the United States does not enjoy politically secure military access ashore.[21]

In important respects this revolutionary view has become a basic element in the Reagan administration's outlook. In September 1981, Weinberger indicated that the "threat" in the NATO area and Northeast Asia seemed less urgent than that in Southwest Asia, which necessitated new priorities and emphases in American security policy. This shift in attitude may come to constitute the most important development in U.S.-European security relations since the early 1950s when the United States first agreed to station forces in Europe for other than postwar garrison duty.

---

[19] *Soviet Challenge*, p. 30.

[20] Congressional Budget Office, *Rapid Deployment Forces: Policy and Budgetary Implications* (Washington, DC: Government Printing Office, 1983), pp. 24–26, and passim.

[21] Jeffrey Record, "Beyond NATO: New Military Directions for the United States," in *U.S. Strategy at the Crossroads: Two Views*, ed. Jeffrey Record and Robert J. Hanks (Washington, DC: Institute for Foreign Policy Analysis, 1982), p. 1.

Second, whether or not it is possible to understand the Mediterranean in its totality, and however independently of Soviet actions, it is necessary to understand Soviet policy and perspective in themselves. Although this may seem a truism, current literature and discussion amply confirm the need for more hard work on this. Given the military risks inherent in East-West competition, both in Europe and beyond it, the need for the "fullest possible information on the mission, capabilities, and deployments of the Soviet Union's Armed Forces" goes almost without saying.[22] But such information must be joined to political perspectives, Soviet as well as Western, more reliably and with more sophistication than generally has been the case in recent years.

In the Mediterranean area, Soviet policy and actions for the foreseeable future are likely to continue to be strategically defensive but tactically and politically offensive. The Soviet Union must continue to meet politically the strategic needs that it cannot satisfy militarily. In peace and crisis Soviet military operations short of war will remain important instruments in Soviet efforts to prevent reassertion of Western dominance in former colonial areas and to reduce NATO's capacity for collective political decisions and concerted military actions. Because the Soviets are merely the beneficiaries and not the primary source of regional volatility, the West will only play to Soviet strength if it insists on defining interests in terms of "stability." George F. Kennan offered wisdom, if not guidance, on devising policy in a world of unbridled change. In 1959 he told the Senate Foreign Relations Committee:

> The task of statesmanship is to be on top of this process, to be influencing it, to have some chance of affecting it by taking the initiative . . . rather than latching onto a status quo that suits you and saying that anybody who tries to change this will meet with your opposition by force of arms. That is no way to try to hold the world together. Constructive planning should guide us not only in meeting the crises of the future, but in doing what we can to shape that future.[23]

Finally, Soviet activity, posture, and goals in the Mediterranean area provide no key to a "vision of the Mediterranean in its totality." Bodies of water, merely by virtue of having a name, give the false impression that their littoral states have more in common than they actually do. Moreover, the historian and policy analyst, through force of intellect, must simplify and attribute meaning to the rude chaos of life and politics. However necessary it may be to order reality, it is dangerous to believe unreservedly that what is plausible is also true. There is no need to postulate the relentless Soviet pursuit, step by step, of some master plan. It is enough to recognize that the interests and methods of the Soviet Union in the Mediterranean, as elsewhere, discommode where they do not outrightly threaten the principles, interests,

---

[22]*Soviet Military Power 1983*, p. 5.

[23]U.S., Congress, Senate, Senate Foreign Relations Committee, *Informal Meeting with G. F. Kennan*, 86th Cong., 1st sess., 12 May 1959, pp. 13, 14.

and institutions of Western democratic states. Soviet political and military roles in the Mediterranean do not and cannot weld the region's diverse problems into a coherent concern for NATO.

The former ambassador to Sri Lanka, Christopher Van Hollen, observes that the prospects in the Persian Gulf seem equally applicable to the wider area under discussion here: "The most likely challenges to Western interests will come from wars between regional states, transborder incursions, civil disturbances, oil embargoes or production cuts, or the overthrow of existing regimes."[24] In much of this, the Soviet Union will play a role, but the Mediterranean, after all, remains a geographical expression and not a political, economic, or even strategic unit.

[24]"Don't Engulf the Gulf," *Foreign Affairs* 59, nos. 4, 5 (Summer 1981): 1065.

THE NATO SOUTHERN FLANK

NATO's Southern Flank: Air Forces

ANTONIO PELLICCIA

From the perspective of NATO air power there are a number of serious problems in the Mediterranean, one of which stems from the domination of SHAPE by the ground forces and by army concepts. The very term "front" suggests a line geographically determined and subdivided into a center and two flanks, each corresponding to the axes of ground operations. But airspace, unlike land space, cannot be geographically confined to serve as a barrier to the enemy's advance, nor does the concept of flanks apply to air warfare. Speed and flexibility of aircraft render lines meaningless in the outer atmosphere. Hence a major problem of all air force command structures has been the need to fit them into an outmoded organizational system, modeled after army and navy commands. It is worth noting that, with the exception of General Lauris Norstad, every Supreme Allied Commander and most of his deputies have been drawn from allied armies, not air forces.

THE NEW IMPORTANCE OF THE SOUTHERN FLANK

These realities, both old and new, should alert Western planners to give greater importance to the theater of operations designated in NATO planning as the southern flank, always considered secondary to the central front which remains the center of gravity for allied forces since Central Europe is still held to be the objective of Soviet expansionism. Despite this assumption, historic experience and present political realities suggest that the Soviet drive in that direction is exhausted. This invites a new appreciation of the situation, one that is more in accord with actual political-military events in Europe and with the Soviet conception of war in general and air warfare in particular.

The importance of Central Europe for military planning is a direct consequence of the political situation that arose at the end of World War II. The principal problem was foreseen by Walter Lippmann precisely on the basis of his knowledge of Russian strategic needs and as a consequence of the neglect by the victorious states to provide certain specific conditions in their definition of a postwar political order. These conditions included a universally accepted definition of the territorial frontiers existing between the Soviet Union and Atlantic states. For Lippmann these frontiers began in the north in Finland, crossed Poland, the Danubian and Balkan nations, and extended to Turkey. All these "border" states, including Sweden, Germany, and Italy, would have to practice strict neutrality in their relations with the only two great powers emerging from the Second World War: the United States and the Soviet Union. To fail to do this and instead to encourage "the nations of Central and Eastern Europe to organize themselves as a barrier against Russia would be to make a commitment that the United States could not carry out," Lippmann warned. The United States was too distant to assure armed support to those nations.[1]

In part these predictions came to pass in the postwar era. To prevent the formation of such barriers, the Soviets did not content themselves merely with the territorial advantages obtained at Yalta; they turned the states of Eastern Europe into satellites and installed Soviet-style systems to assure their control. Soviet expansion toward the West in this sense went beyond the traditional geographic dimensions of the tsarist empire. These dimensions were first suggested by Ivan the Great who between 1462 and 1505 identified three paths of expansion: the Baltic Sea, the Black Sea, and Siberia; his successors pursued, tenaciously and with considerable success, the policies of territorial aggrandizement. Ultimately, Catherine the Great succeeded in conquering the Crimea, the northern coast of the Black Sea, and part of Poland through a partition of that state with Prussia and Austria. Catherine did not, however, resolve a resulting problem, that of access to open water through the Dardanelles. When seen in its historic setting, and considering the strategic and economic requirements of the Soviet Union in the Persian Gulf and Indian Ocean, Moscow's postwar political design in Europe was clear: having attained a certain stability in northern and Central Europe, it must orient its efforts toward the south in order to complete Stalin's work at Yalta. Blocked by early Western postwar policy, this whole Soviet strategy has received new impetus in the last decade.

John Erickson arrived at a similar conclusion after studying the problem exclusively from the military point of view. In fact, he has written that the two flanks, north and south, can no longer be considered secondary to the central front, both because the region has reached a certain degree of political stability, and because the Soviets have achieved superiority in conventional arms. This situation has induced increased Soviet attention to the north and south in order to prepare and carry out a grand pincer movement that would facilitate the fall of the central front.[2] The Soviets have divided the entire geostrategic area in which they are interested into "theaters of military operations," coordinating them at the highest political-military level. Soviet policy

[1]Walter Lippmann, *U.S. Foreign Policy: The Shield of the Republic* (Boston: Little, Brown & Company, 1943), p. 149.

[2]John Erickson, "Soviet Options and the Southern Flank," Erickson lecture notes (University of Edinburgh, 1978), p. 13.

for the southern theater of operations, involving massive qualitative and quantitative air power, can only be evaluated properly in relation to the aim of Soviet leaders to secure one of their ultimate historical territorial objectives, the Straits. To this end the Soviet Union is attempting to consolidate its presence in the Mediterranean as a great naval power and to reduce the influence of Western nations, above all the United States.

THE SOVIET CHALLENGE

In its military contingency planning for southern Europe, the Soviet Union has massive numbers of modern aircraft available for this theater. The majority of these are interceptors and fighter-bombers, including those stationed on two aircraft carriers in the Mediterranean, but there are also bombers assigned to bases in the Crimea. Soviet military aviation in recent years has undergone a substantial transformation in the quality and quantity of its first-line aircraft by increasing their payloads, armaments, avionics, and general performance. From a doctrinal point of view, Soviet aviation applies Giulio Douhet's fundamental principle of aeronautical unity and assigns aerial dominance, or its intermediate steps—superiority, supremacy, predominance—as the strategic goal of its air forces through the destruction or neutralization of enemy aviation. The Soviet command organization and order of battle confirm this. Moreover, Air Marshal P. S. Kutakhov lists air operations in this order of priority: operations to destroy enemy air forces and gain aerial supremacy, operations to destroy operational reserves or lines of communications, and operations to destroy military industries.[3]

By centering command at the highest level, the Soviets aim at the maximum flexible utilization of their air resources and at the concentration of their air forces at decisive points. To achieve this goal, Soviet aviation is divided into five components: air defense forces, long-range, frontal, transportation, and naval. The air defense forces utilize, with equally varied armaments and avionics, about 2,200 aircraft of various sorts—Su-15s, MiG-23s and MIG-25s, Yak-28Ps, as well as Tu-126 airborne warning and control systems (AWACs). Long-range aviation consists of about 900 aircraft including almost 100 Tu-26 *Backfires*, while frontal aviation includes over 5,000 first-line combat aircraft, with 3,000 in reserve. Transport aviation possesses about 1,800 An-12 and An-22 aircraft and 2,600 helicopters, while naval aviation utilizes about 800 aircraft including approximately 50 Tu-26s and about 450 helicopters. In conformity with the aerial doctrine noted above, the Soviets assume an in-depth offensive role by employing the Su-24 *Fencer*

---

[3]P. S. Kutakhov, *The Conduct of Air Operations: Selected Soviet Military Writings* (Washington, DC: Government Printing Office, 1977), p. 240.

with its terrain avoidance and range-finding radar and by adopting a task force organization similar to that employed by the United States. The Warsaw Pact provides an additional 2,500 aircraft of all types including the most modern Soviet airplanes such as the *Flogger B* and *D* fighter-bombers, stationed in Czechoslovakia and East Germany, that can attack many objectives in northern Italy.

In addition to these forces, the Soviets have an imposing number of reconnaissance, electronic warfare, and refueling aircraft as well as hundreds of planes in the hands of pro-Soviet Arab states. Of particular importance is the Libyan air force, equipped with 46 *Mirage* F1s, 58 *Mirage* 5s, 161 MiG-23 *Floggers*, 50 MiG-25 *Foxbats*, 7 Tu-22 *Blinders*, land-based air defense with early warning radar lines and Soviet-supplied SAM 2s and SAM 3s, and well-protected airfields with SAM 6s and French-supplied SAM Matra *Crotales* for low-flying attacks. The Libyans also currently have more than 100 combat aircraft on order from a number of suppliers. Several bases and many landing strips prepared by the Soviets in the desert can simultaneously handle combat and transportation aircraft.

NATO AIR DEFENSE

Overall it is evident that the aerial threat on the southern flank is assuming daily more menacing proportions and making the defensive problem of the allied nations in the area, especially Italy, increasingly complicated. NATO opposes this threat with about 950 aircraft of various types and specialties assigned by Italy, Turkey, Greece, and the United States to the 5th Allied Tactical Air Force (ATAF) headquartered at Vicenza under an Italian commander, the 6th ATAF headquartered at Izmir under a Turkish commander, and the 28th Greek TAF at Larissa. These forces are under the Commander Allied Air Forces Southern Europe (COMAIRSOUTH), an American general who operates from Naples at the site of the Air Forces Southern Europe (AFSOUTH) high command. In case of war the U.S. Sixth Fleet would come under the operational command of the Commander in Chief Allied Forces Southern Europe (CINCSOUTH), and thus the approximately 200 aircraft on the two aircraft carriers must be included in the southern European order of battle. In addition, NATO plans contemplate five contingency areas in which reinforcing aircraft can arrive as needed, air defense on land is provided by *Nike* and *Hawk* missiles.

Recently the Turkish, Greek, and Italian air force chiefs of staff have expressed strong concern over future deficiencies in the air defense of the southern flank. Because of problems arising from their geographic position, these three nations have military requirements that do not always coincide with NATO's. Moreover, unlike the air forces of Central Europe under the direct control of Allied Air Forces Central Europe (AAFCE), which has achieved centralized command of all aerial operations on the central front, the air forces of southern Europe operate regionally. Organizational and operational problems that have been resolved, at least in part in the center, are aggravated by the impact of important political issues in the south.

In addition, a number of doctrinal differences between U.S. and European air forces threaten to produce serious confusion in the area of employment, the most obvious example being the definition of counteraviation as a tactical operation in air-land battle strategy. The Europeans also persist in trying to protect their own national identities and the capacity to employ their air forces independently of NATO in case the complicated alliance consultation and decision-making mechanism is slow to react to a surprise attack. Moreover, the different flight characteristics and capabilities of U.S. and European aircraft impose varying tactical employments and at times opposing operational concepts. Debate continues between supporters of point defense and those of area or front defense.

Finally, the major problems of coordinating air defense with antiair and counterair defense and the substitution of the Identification Friend or Foe (IFF) with the more trustworthy NATO Identification System (NIS) remain unresolved. In this regard the recent conflicts in the Middle East and Falkland Islands have needed the divergence of view between Americans and Europeans over the utility of employing sophisticated technology, sustained by the former, as opposed to the faith placed by the latter in human capacity. The outcome of these conflicts showed the Americans to have been correct; the Europeans have abandoned their reservations about the AWACs, the Joint Tactical Distribution Information System (JTDIS), and the development of all four techniques of electronic warfare—standoff, self-screening, close support, and ECM escort.

About two years ago the NATO Air Defense Committee (NADC) was created to resolve these problems. This organization is charged with studying and coordinating among all the allies a general and realistic plan of air defense. Among the many advantages that it is hoped NADC will provide are discussions on air defense planning for France, Greece, and soon Spain. All the allied air chiefs of staff are in agreement on the utility of this new organization. However, as the Italian air chief of staff has observed, it remains to be seen if this committee will resist the inherent danger in organizations of this type and level and not lose its sense of reality, particularly with regard to the limited financial resources of most of the allied nations. The fact also remains that NADC is only a consultative study group and as such cannot resolve the three fundamental problems of the air forces: control and coordination of all aerial operations at the highest possible level, rapid passage of national forces into the NATO command structure in case of war, and, finally, strict coordination between air and surface force commands.

It may be useful to point out that Italy's military structure is so completely identified with NATO that operational plans, the composition and deployment of forces, their interforce balances and budgetary outlays, and annual exercises are all conceived and carried out in relation to this identification. The Italian air force in particular has a reorganization in progress in which the command of all air forces in the region will be exercised by the 5th ATAF.

Considering the political, economic, and social problems of the southeastern Mediterranean nations of the alliance, the doctrinal divergences, and the organizational and technical inadequacies in NATO's southern flank, a wide gap exists through which the air forces of the Soviet Union and its allies can penetrate to the Mediterranean. It is time to close that gap.

NATO's Southern Flank: Land Forces

ALBERTO LI GOBBI

With only a quick glance at a map, it can be concluded: a) that the strategic problems of the southern flank can be identified with the geostrategic problems of the Mediterranean; b) that the Southern Region is much larger and more diverse in its geographic, political, ethnic, and historic characteristics than the other NATO regions; c) that the lack of territorial contiguity among NATO member states will make mutual assistance and planning for external support from the United States, England, and Portugal impossible unless NATO maintains control of the Mediterranean; and d) that domination of Mediterranean sea lanes will be irremediably compromised if one of the five southern flank states were defeated on land and occupied by the enemy. All of the foregoing problems apply particularly to Italy, whose territory is located in the heart of the Mediterranean.

In sum, NATO's southern flank is an extremely large and complex operational theater in which very close coordination among land, air, and sea operations is of vital importance. Like modern *curias,** land forces are divided by enormous distances and scattered over widely varying terrain. The outcome of a land battle in any one of the three *curias* (Italy, Greece, or Turkey) inevitably would be unfortunate without air and naval control of the Mediterranean. Likewise, air and sea control of the Mediterranean would become practically impossible if one or more of the three *curias*, particularly Italy, were defeated and invaded.

Having sketched the broad lines of the geostrategic operational characteristics in the Southern Region, we can now pass to the various individual factors that create the southern flank problem. The first of these elements is the Mediterranean, an enclosed sea with only three entrances—Gibraltar, Suez, and the Dardanelles—and with a number of internal choke points. In peacetime about 5,000 ships of medium tonnage and 1,500 transoceanic ships of large tonnage cross the Mediterannean each day. This total includes approximately one-third of the tankers that carry essential petroleum products to France, Germany, and Turkey and all the oil destined for Greece, Italy, and Switzerland. Among these ships a daily average of 150 are Soviet and transport an average of 50 percent of Soviet imports and over 60 percent of that nation's exports.

---

*A reference to the subdivisions within the three ancient Roman tribes.

NATO's MEDITERRANEAN NATIONS

The NATO nations of the southern flank are notoriously weaker from an economic viewpoint than the other member states. Their utilizable military infrastructures, such as roads, railroads, navigable rivers, and port installations, are inferior to those possessed by the central region nations. These twin weaknesses of infrastructure and economy can have only a negative effect on the military instruments of this NATO region. It is against this background of economic and infrastructural difficulties that the three *curias* of the south can be individually reviewed.

Italy

The Italians have changed governments more than forty times since the end of the Second World War. All of these governments have solemnly proclaimed their unwavering loyalty to NATO, both vocally and in print, but their deeds have been less impressive. For example, the progressive decline of the percentage of public expenses devoted to defense, which fell from 15.47 in 1966 to 9.6 in 1975 and to 5.1 in 1981, does not suggest an optimistic outlook for the future. Similar nations within and outside NATO—France, Germany, Switzerland, Greece, and Turkey—for years consistently have maintained an average of 20 percent investment in defense. The dangerous Italian decline in support has taken place despite the solemn 1979 undertakings of all NATO leaders, including Italy's, to augment annually defense budgets in real terms by 3 percent over inflation.

The surprise military reorganization of 1975–76, without prior consultations within NATO, already has reduced by two-thirds the size of Italian forces under NATO control. Since then both low-level training and operational readiness, due to lack of resources and the pressure of events, has plagued the Italian armed forces, especially land forces. In spite of this the recommendations made in the NATO annual review are always ignored. The most recent "reduction through reorganization" of the military has brought it to a level barely adequate for continued survival even for peacetime purposes. Reduction through reorganization, which at least for now is meeting only meek opposition from the defense community, seems to have been decided upon without the usual prior NATO consultation. The one positive fact in such a negative landscape is the agreement to put missiles in Comiso. This decision, however, remains strictly on paper for the present and is still vulnerable to eventual rethinking and last-minute reversal.

Greece

Andreas Papandreou's campaign platform included the withdrawal from NATO and the European Economic Community, the removal of U.S. bases from Greece and of all fleets from the Mediterranean, and the abolition of

NATO and the Warsaw Pact. At this point it is not clear if this was simply empty oratory since until now nothing substantial has emerged from all of this. Greece is an essential element in the defense of the Mediterranean; its permanent loss to the NATO alliance would represent a formidable triumph for Soviet foreign policy.

Turkey

Turkey borders directly on the Soviet Union as well as on Bulgaria. In addition it has neighbor states—Syria, Iran, and Iraq—that, while not Warsaw Pact members, are anything but pacific. Turkey's internal stability is an indispensable element for the defense of NATO's southern flank, and its military government has helped promote that stability. The price of military government, however, remains high. The absence of parliamentary liberties separates Turkey from its allies, and its collaboration in the Turkish Cypriot attempt in 1983 to carve an autonomous, if not fully independent, republic on that island further exacerbated its contentious relations with Greece.

Spain

The Spanish armed forces are traditionally organized and oriented toward territorial defense. Its army comprises 240,000 men, only 150,000 of whom are fifteen-month draftees. Thus, it is more numerous than the present Italian army of 180,000, whose total complement of 200,000 men are twelve-month draftees. Moreover, due to the long-term service of its draftees, presently fifteen instead of twelve months, the Spanish army can take advantage of an operational training program which is superior to that of Italy. The Spanish navy, although very small, is undergoing reorganization and already possesses a high degree of training and a notable surface and underwater operational capacity. Spanish aviation, with its more than 190 combat aircraft, has a good operational capacity, especially in air defense and close support for ground operations.

In his inaugural address, Felipe González asserted that he would honor his campaign pledge to put the issue of Spanish membership in NATO to a national referendum. In effect, however, he has placed a final decision in the deep freeze. While awaiting the outcome of an eventual referendum, the process of planning for Spanish involvement in NATO's southern flank continues.

France

Even after France's withdrawal from NATO's integrated military structure, cooperation with the French has remained active and, with the election of François Mitterrand, has even improved. France has a large naval force in the Mediterranean, including two aircraft carriers. A memorandum of cooperation between NATO and France permits and facilitates their by now traditional joint participation in naval exercises. In cases of conflict, the French fleet could provide a notable contribution to an air and naval victory in the Mediterranean, and therefore to the favorable outcome of the battles in the three land *curias*.

THE SOUTHERN REGION'S GROUND THEATERS

Even a summary examination of single factors and operational conditions immediately reveals that peace in the Mediterranean, the heart of NATO's Southern Region, is threatened not only by the Warsaw Pact states but also directly or indirectly from almost any point on the horizon. From the classic doctrinal viewpoint, four potential theaters exist on NATO's southern flank, three of which are ground theaters and the fourth—the Mediterranean—is an air and naval theater imposed among the others to safeguard their connecting links. Although the latter theater will be bypassed in the following discussion because it is extraneous to this subject, it remains an essential element for the comprehension of each of the three primarily ground theaters of operation. The one factor common to all three of these land theaters, as was the eventual case of the ancient *curias*, is that the Warsaw Pact forces, which have the initiative, are numerically superior and possess more modern equipment for fighting an air-land battle. Moreover, they also are better equipped and trained for the eventual use of chemical warfare.

The Italian Chessboard

Italy constitutes the left wing of the defensive arrangement for the Southern Region. Its conquest by an enemy, besides securing the resources of an area with an advanced industrial base and physically separating the southern from the central regions, would also permit the forward deployment of Warsaw Pact aircraft on the airfields of northern Italy. This would have two immediate results: to increase the threat to southern Germany and eastern France and to extend the radius of action of Warsaw Pact air forces up to the Straits of Gibraltar and the entire North African littoral, with a decisive impact on the outcome of the naval air battle in the Mediterranean.

The Italian chessboard thus can be subdivided into two regions with morphologically distinct characteristics. First, there are its mountainous parts characterized by an impervious, highly compartmentalized countryside in which the possibilities for movement, especially of mechanized and armored units, are confined to the lower valleys. Second, the plains coincident with the Friuli-Venezia Plain constitute the natural route of attack from the east through the plains of Gorizia toward the Po Valley. As for possible starting points for an attack from that direction, the most dangerous is doubtless from the Kranj-Lubljana shell following the course of the Vipac to the Gorizia plains. The characteristics of the terrain justify the adoption of operational plans tied to the concept of forward defense which, in addition to constituting NATO's official doctrine, finds an ample tactical and strategic justification in this area. In fact, only after the eventual conquest of carefully prepared forward positions can the enemy deploy its forces, taking full advantage of its numerical superiority.

Recent NATO estimates place at seventeen the number of Warsaw Pact divisions capable of operating on the Italian front. The spearhead of these forces consists of its armored units which include 4,900 tanks. In both quantitative and qualitative terms this threat is constantly increasing. Ongoing modernization programs undertaken by the USSR and Hungary include the updating of tank forces, with the introduction of new models in place of obsolescent ones, an increase in the firepower and range of tactical missile units, and a considerable improvement in the quality and quantity of attack helicopter units. On the other hand, NATO's land forces on the Italian front presently consist of twenty brigades (eight mechanized, five armored, five alpine, and two motorized) plus one missile brigade. These forces, all of which are Italian, are organized into three corps: the 5th (mechanized) is responsible for the defense of the plains, the 4th (Alpine) for mountain defense, and the 3d, with its mechanized and motorized units, constitutes the reserve.

NATO's forces on this front are considerably inferior to those of the Warsaw Pact, about 1 to 3. As a consequence, NATO can only oppose an invasion from the east through a preplanned and speedy utilization of terrain especially fitting the needs of the defenders. Fortunately, in view of the ratio of military forces, the geopolitical situation is presently favorable to NATO. Austria and Yugoslavia are awake to the dangers of their buffer status and would provide a notable warning time for the allies. Still, NATO forces in the Italian sector do not enjoy truly adequate reserves, an efficient mobilization scheme, or fully effective command and control systems capable of exploiting an early warning; that is, they could find it difficult to exploit the usable warning time in order to permit external reinforcement.

The Greece-Turkey-Balkan Front

On the Balkan front the Warsaw Pact enjoys a 2-to-1 air and land forces advantage; the threat comes from Hungarian, Romanian, Bulgarian, and Soviet forces. Northern Greece consists of about 1,000km. of borders, with half of this defensive area having a very limited rear, ranging in depth from 25km. to 75km. The terrain here is mountainous, running north and south, and the valleys, with rivers running parallel, constitute natural lines of communications. The most important and natural direction of attack—the traditional path of invasion of Greece—is from Yugoslavia. Ultimately, these were the routes followed in both world wars. A successful invasion in this sector would cut off part of Greece and all of Turkey from the rest of the alliance. Moreover, the limited depth of southern Greece and the relative openness of its terrain to the north would work against an energetic and tenacious defense.

On the far east of the Balkan front, the terrain is quite different. Turkish Thrace primarily consists of open undulating terrain with sparce vegetation and excellent visibility. The Turkish Straits constitute an obvious primary objective for Warsaw Pact armored units. Without control of the Bosphorus and Dardanelles, the Soviet Black Sea fleet cannot reinforce its Mediterranean squadron or sustain naval operations in the Mediterranean, except for a very brief period. Therefore, a major Soviet land effort to secure these passageways can be predicted. The Turkish 1st Army is charged with the defense of Thrace, but it has only a few tanks, limited antiarmor capacity, and inefficient air defense. It would be hardpressed to defend itself for any long period against a heavy armored attack.

Eastern and Southern Turkish Fronts

Bordering the Soviet Union for 600km. and Iran, Iraq, and Syria for a farther 1,700km., eastern Turkey is perhaps the least known zone among the operational theaters of the Southern Region. This area is particularly important because it blocks Soviet access to the Persian Gulf. However, the Soviets have forces on the Transcaucasian front that are greatly in excess of reasonable deterrent needs, and thus they enjoy a 7-to-1 superiority. Turkish forces on this front are not capable of prolonged operations and seriously need modernization. On the other hand, eastern Turkey is mountainous, underdeveloped, and subject to extreme temperature variations, all of which favor the defender. Moreover, its enormous distances permit sacrificing territory to gain time. In the south small Turkish forces have to face a large number of Syrian and Iraqi divisions, and, although the efficiency of these nations' forces is open to discussion, the entire border with Syria and Iraq lies open and is difficult to defend.

In summary, the commander of allied forces in southeast Europe has enormous problems to resolve and difficult decisions to make in order to defend the common borders with Bulgaria and the USSR and to oppose a possible worst case coordinated attack from Syria and Iraq in addition to defending 1,600km. of Black Sea coast. However, it is true that Soviet forces in this region are not directed primarily at Turkey but are aimed at the south, with the mission of interdicting Europe's principal oil arteries in the Persian Gulf. In this situation, Turkey would serve as an efficient deterrent, menacing the right flank of an invader.

CONCLUSION

The Southern Region, or southern flank, is also called NATO's Achilles' heel because of its vast size, complexity, and grave problems which threaten world peace today; the Arab-Israeli conflict and the oil crisis are now added to the traditional Russian drive to the sea. A NATO defeat on any front of the southern flank could lead to the loss of control of the Mediterranean and be a fatal blow to the entire alliance, including the central region. Simply stated, the entire Southern Region can be defined in its totality as both an extremely sensitive front—one that is difficult to defend—and important, as its loss would be serious, if not catastrophic.

Although the Italian front is not particularly sensitive at first glance, given the buffer states around it and the nature of its terrain, it is still vitally important to NATO's role in the Mediterranean. In fact, it is so central to the alliance's survival in the area that weakness and disarray of NATO forces could induce an enemy to strike here despite the inherent difficulties in attacking that particular *curia*. Italy's loss inevitably would carry with it NATO's surrender of control over the Mediterranean and the separation of the alliance into two parts.

In light of this scenario, it is difficult to explain both the political insensitivity, on the part of the Italians, toward NATO in general and the military insensitivity toward the gap on the Gorizia plain in particular. As a case in point the percentage of disposable public funds scheduled for defense purposes has risen by only one-fourth of the defense commitments of similar and even poorer allied states. Instead of increasing and revitalizing their armed forces, Italian leaders are thinking of further cuts in their military establishment. One must ask if it is possible that Machiavelli's political descendants have understood that Italy is too important to be overlooked by an enemy aiming at control of the Mediterranean. Today, at one of the most dangerous periods in history, Italians seem to have decided to leave the duty and honor of defending their homeland to their allies.

Mediterranean Naval Forces

MARIANO GABRIELE

It may fairly be claimed that today the Mediterranean is the most crucial maritime area in the world.[1] As has been amply documented in preceding essays, this was not true in the earliest years of NATO. In that first stage, the Mediterranean was on the periphery. Allied strategic scenarios assumed Central Europe to be the critical focus of the organization. It was there, after the end of World War II, that Soviet domination of Eastern Europe and the emergence of the German problem generated much of the tension between East and West. The strategic geography of the time pivoted on that part of the East-West contest; the northern and southern flanks of NATO were simply distant wings of the more vital central theater.

In these postwar years the Mediterranean seemed like a western lake, the uncontested dominion of the Anglo-Saxon powers.[2] From Gibraltar to the Dardanelles, action on the NATO southern flank was to be coordinated with the main action on the central front in the event of war. Although since November 1943, at the Tehran Conference, the Soviets had asked for a naval base either in the Marmara or Aegean seas, they had accepted without much agitation the Western rejection of this request as well as their demands to revise the Montreux Convention. In the same way, Molotov's request had no success in September 1945 at the London Conference, in which the Soviet Union asked for administrative control over Tripolitania (Libya) so that it could "occupy a seat it fits and . . . in this way, to have some bases for the mercantile fleet in the Mediterranean."[3]

Even in Stalin's last years, however, it was possible to anticipate Mediterranean difficulties in the near future. The traditional Russian frustration over restricted access to the major world oceans had led to a renewal of naval construction and a revival of Soviet interest in its navy as early as the 1950s.

---

[1] I. B. Ande, "El rol del poder político y militár en el Mediterraneo," Toulouse conference, France, 1–4 October 1980.

[2] M. Gabriele, "L'uso politico della forza militare in Mediterraneo," *Rivista Marittima* 111, no. 6 (June 1978): 27–33; F. Micali-Baratelli, "La sicurezza del Mediterraneo e delle vie marittime occidentali ed i limiti dell'area atlantica," *L'alleanza atlantica e la difesa dell'Europa* (Rome, Editrice Europea, 1978), pp. 79–115.

[3] M. Gabriele, "Mediterraneo (1945–1953)," *Rivista di studi politici internazionali* 46, no. 1 (January–March 1979): 25–48; R. G. Weinland, "Soviet Strategy and the Objectives of Their Naval Presence in the Mediterranean" (Paper presented at IAI conference on "Growing Economic Interdependence and the Future of Security in the Mediterranean," Castelgandolfo, Italy, 6–9 September 1982).

Perhaps, the Korean lesson had some effect. It had been a lesson that had shown how a strong naval power suddenly could consolidate and extend amphibious successes, taking the war very far from the front and behind the enemy.

Soviet attention to the sea, the Mediterranean in particular, increased little by little after the first Arab-Israeli war in 1948. Following the 1956 conflict, Moscow announced that it would rearm Egypt and became more assertive in the area. Although Western influence was still strong as far as existing forces were concerned, there was some softening of the status quo. The Soviets were emboldened, both by the anti-Western stance of Nasser and the Western debacle at Suez, to renew their hopes for a presence in the Mediterranean.

After 1956 it seemed to Moscow that new opportunities were emerging. Although until then the American policy of containment had dissuaded them from taking major initiatives in the Mediterranean, the Soviets thought it possible to undertake small projects that also were destined to produce important changes in the long run.[4] Their objective at the beginning of the 1950s was not to present a threat to NATO's southern flank but to claim an interest in and a right to exercise influence in the Mediterranean. In April 1946 the Italian ambassador in Moscow identified a constant element in Soviet interest in Mediterranean politics; namely, that the USSR tended to expand its influence whenever and wherever it was possible, without necessarily expecting victory on every occasion. In such a contest from Trieste to the Straits, from the Dodecanese to Libya, the eastern Mediterranean became a setting for actions in support of Soviet expansionist ambitions.

In brief, a series of phases of Soviet activities in the Mediterranean can be identified. The first is a clear Western predominance placed in the period from the end of the war to 1956. A second phase covers the years from 1957 to 1966, during which time the Soviet presence and influence increased in the Mediterranean and in the world, with the construction of great fleets that could challenge the status quo on the high seas. This phase gave way to the first concrete threat to the southern flank of the Atlantic Alliance as crises developed in the Aegean and in the Middle East. The Soviet navy was the instrument of confrontation, managing to appear on the scene as a source and symbol of Soviet power. While the United States was engaged in Asia and in the Caribbean and other European powers—France, Great Britain, and Belgium—in Africa, new local pressures were exerted on NATO's southern flank. Still, to the West, Soviet activities seemed a diversion from the main problems of the central sector. Western superiority in the Mediterranean was not yet in danger.

Little crises were exploited by the Soviets as part of their larger plan of destabilization of the area. The Straits and the Suez Canal took on increasing

[4]L. Garruccio, "Mediterraneo e multipolarismo," *Affari Esteri* 3, no. 35 (July 1977): 413–25.

importance for Moscow in the context of its new global strategy, but for the most part its strategy was based on a general condition of peace in which NATO could localize conflicts and the challenge could be minimized. The main difficulty over the exclusion of the Soviet Union from the Mediterranean was not really a juridical one. The critical issue lay in the Soviet objective of using its naval power to control the Straits and the Suez Canal. It sought positions—army, navy, and air—that would promote interlocking bases.[5] In this scenario the Soviet Fifth *Eskadra* played a special role in applying pressure on Turkey and Egypt.

Despite Nikita Khrushchev's repudiation of Stalin's behavior toward Turkey and despite his support of Nasser over the nationalization of the Suez Canal, neither country had fulfilled Soviet expectations.[6] Turkey remained impervious to both flattery and threat, while Nasser insisted on maintaining substantial autonomy in his relations with his Soviet patron. As long as the keys to the doorway of the Straits remained in possession of an Atlantic power, the utility of support points in the Mediterranean would always be somewhat tenuous.

SOVIET TACTICS IN THE 1970s

The Arab-Israeli Six-Day War marked a turning point. Although the Soviet navy already had begun a cautious expansion in the Mediterranean after the war, it registered a remarkable increase in its presence there, reaching a peak at the time of the fourth Arab-Israeli war in 1973. After Nasser's death in 1970, however, Nasserism and Arab nationalism worked against Moscow. In spite of an Egyptian-Soviet treaty of friendship and cooperation, the Soviet advisers were expelled from Egypt in 1972 and the treaty was denounced in 1976. The accords of Camp David followed. The role of Moscow declined from manager to spectator of the crises in the East.

While Soviet power seems to have receded in the past ten years, appearances can be deceptive. Actually, during the 1970s the Soviets conducted pincer movements of remarkable political and military importance. Just as they appeared to be expelled from the area, they began to expand their activities on the southern and eastern flanks of the Mediterranean. In Africa, using Cuban troops, along with Warsaw Pact forces, they moved over to the Atlantic coast of Angola and to the conjunction of the Red and Arabian seas

---

[5]H. Labrousse, "Sécurité en Méditerranée," originally presented at Monaco conference, 19–21 February 1981 and subsequently published in *Défense Nationale* 37, no. 6 (May 1981): 65–77.

[6]W. Berner, "La politica sovietica di alleanze nel Mediterraneo e la distensione," in S. Sivestri, *L'uso politico della forza militare in Mediterraneo* (Bologna: Il Mulino, 1977), pp. 87–99.

in Southern Yemen, where they already had a foothold. In the Middle East they took advantage of the Iranian revolution to invade Afghanistan, opening a new threat to the Persian Gulf.

These operations changed general strategic conditions radically. The Mediterranean Sea was transformed, in a manner of speaking, from a southern flank of NATO to a new central theater, while the central zone became the northern flank. The Persian Gulf by the end of the 1970s suddenly emerged as a kind of new southern front, even if that fact was not fully clear to the NATO allies.[7] Such a front with its epicenter in the Mediterranean holds many advantages for the Soviet Union. Among them is the political nature of many of the conflicts in the area in which the Soviets could exploit the anticolonial sentiments of Third World nations against the West. Moreover, it is a mobile front rather than a fixed one like the eastern borders of the German Federal Republic, with its open spaces for maneuvers. If one accepts the idea that neither superpower wants a hot war, with its catastrophic consequences, it is surely true that the northern wing is the steadiest and the quietest, with its well-defined positions and reciprocal awareness of the disastrous results that could follow rash political and military movements.

In the Mediterranean area, on the other hand, the borders are unstable and uncertain. The eastern and northern Africa regions, which are included in this context, suffer from political, social, and economic conditions that favor foreign interference in their domestic politics. Rebellions, political intrigue, social and economic crises, feuds, and genocides abound in the area. Destructive internal ferocities occur, such as the tragedy of Lebanon demonstrates, in which local conflicts lead to long, unquiet armistices that cannot substitute for peace.[8] Such situations are made to order for foreign manipulation.

The main problem of the NATO southern flank is that the wide scope of potential conflict in the area does not coincide with the geographical boundaries of NATO. As an alliance, NATO cannot respond to challenges outside the zones defined by the North Atlantic Treaty in 1949. The United States can—and has—acted outside these boundaries, as have France and Britain on occasion, but, while they may serve NATO interests, they function as individual powers rather than members of an alliance. It is useful to underline the fact that single marginal interventions, such as the presence of forces coming from the Atlantic powers in Lebanon, are decided in the context of national and not of NATO politics.

On the opposite side, the use of elements from its Warsaw Pact allies cannot seriously be assumed to result from national decisions but rather flow from Soviet political choices. In such a framework the eastern Mediterranean between the Straits and Suez is an integral part of the southern front for the

---

[7]Micali-Baratelli, "La sicurezza del Mediterraneo,"

[8]M. Gabriele, "Mar Arabico, Oceano Indiano e Mediterraneo," *Rivista Marittima* 111, no. 10 (November 1978): 15–23.

Soviet Union, a front essentially opened in the 1970s. To secure this strategic connection, the Soviets use their Fifth *Eskadra* in the Mediterranean.

An analysis of the Soviet naval presence in the Mediterranean, compared with its naval presence in the world, seems to confirm this assumption. From a naval perspective three different periods may be identified. In the first one (1949–71), the increase in the Soviet naval presence in the Mediterranean passed from 1,500 to 19,000 ship days in one year, with a developing rhythm that was more intense than the corresponding increase among NATO forces. In the second period (1972–76), there was a less rapid increase from 19,400 ship days in 1972 to 20,300 in 1973, the year in which the Yom Kippur War occurred. In the third period (1977–83), there actually has been a decrease in the Soviet maritime presence in the Mediterranean to a lower but steady level: 16,500 ship days in 1977 and since then averaging about 17,000 per year. Data comparing Soviet regional and global naval operations reveal that, while in the first phase the effort sustained in the naval field in the Mediterranean was more than one-half of that for the total world oceans for the Soviet Union, in the third phase, especially in the last five years, it was less than one-third of the total.

Looking at Soviet global naval activity, it is easy to see that, after a pause during the years from 1975 to 1977, increases took place not only in the Mediterranean but also in the Pacific and Indian oceans. The Fifth *Eskadra* remained almost the same. During recent crises in the Middle East, it has included about forty-five units, while during the time of the Yom Kippur War it was at ninety-six units, and this in the presence of great American naval concentrations in the Mediterranean during the 1982 Lebanese crisis. It is clear that the actual size of the Fifth *Eskadra* is considered by Moscow sufficient to sustain its desired role.

An explanation for Soviet restraint may be found in disillusionment over the behavior of its Middle Eastern clients over the past fifteen years, yet the Mediterranean Sea, as noted, has always been a main interest for the Soviets, and the navy remains the main instrument for their penetration into the Third World. Their naval presence has permitted gradual penetration into distant areas, preparing the way for future military involvement. The traditional use of the fleet as an instrument of diplomacy and power projection takes fewer risks than armed military intervention both in the Mediterranean and in the rest of the world.[9]

It seems that Soviet naval power in the Mediterranean has become an extremely important element in the USSR's overall global strategy. Two objectives seem to be entrusted to the Fifth *Eskadra*, as they have to other key elements of Soviet sea power in the world: "the active defense of peace

[9] B. Blechman, "The Scope and Limits of Military Power in Third World Conflict" (Paper presented at International Institute for Strategic Studies, twenty-second annual conference, Stresa, 11–14 September 1980).

and of progress" and the promotion of "progressive change." The first consists in the achievement of a strategic deterrence which is directed to the defense of the USSR as far as possible. The second objective is to be reached through pressure on the status quo. Revolutions and local wars are the norm in turbulent emerging nations, and naval diplomacy is a useful means to influence events, particularly in those countries formerly in the Anglo-American sphere.[10]

The political role is more effective when the navy appears capable and ready. To maintain credibility it is necessary for the Fifth *Eskadra* to train itself in new tasks in case of war. These tasks, in a traditional land power such as the Soviet Union, concern general combat activities, which the navy depends upon more heavily than in the case of Western nations. For instance, it must practice operations ranging from a preventive action against the Sixth Fleet to a strong attack against NATO maritime communications.

PROBLEMS OF NATO's SOUTHERN FLANK

Given the strict geographical limits of the Atlantic Alliance, NATO has self-imposed inhibitions against coping with issues in nations south of the 23d parallel, or to deal with Soviet intervention there. The limits, however, are artificial, and it would make sense for NATO to take into account the results of inaction in areas that may be outside artificial boundaries but that influence the fate of the southern flank.[11]

It has been said above that there is, in fact, a new kind of southern front of the alliance, considering the western interests in the Red Sea and in the Persian Gulf.[12] Two caveats, however, are necessary to interpret this assertion correctly. The first is that it is necessary to eschew any excessive alarmism, avoiding the temptation to speak, for example, of Soviet domination of the Middle East that never existed and does not exist today.[13] The second is that the NATO military structure is only one component of Western resistance to expansionist Soviet pressure. In addition, the new French military doctrine must be recognized. It is clearly European and Mediterranean as well as supportive of French national interest.[14]

---

[10]Weinland, "Soviet Strategy."

[11]Micali-Baratelli, "La sicurezza del Mediterraneo."

[12]R. Aliboni, "Il fianco sud della NATO. La situazione nel Mediterraneo e nell'Asia sudoccidentale. Ruolo americano, e ruolo europeo" (Paper presented at IAI conference on "Negotiations for the Limitation of Nuclear Arms in the European Theatre: Security, European Integration, and the Mediterranean," Rome, February 1982).

[13]E. V. Rostow, "The Giant Still Sleeps," *Orbis* 24, no. 2 (February 1980): 311.

[14]Blechman, "The Scope and Limits." For additional references see Micali-Baratelli, "La sicurezza del Mediterraneo," pp. 97–106; M. Cremasco, "The Mediterranean, the Atlantic and the Indian Ocean: A Difficult Strategic Equation" (Paper presented at Toulouse conference, 1–4 October 1980); and I. Isnard, "Une politique navale pour l'an 2000," *Le Monde*, 14 July 1982, p. 8.

However, there is a different element of critical concern. The image of the West, in the opinion of the Third World, remains conservative. Some believe that the superior quality of Western goods, compared with the inhumanity, inefficiency, and bureaucratization of the USSR, is sufficient to give a decisive advantage to Western influence in the Third World. That is far from true. Objective conditions in many emerging countries engender resentment and envy for the West which is considered "rich" and "diabolic." The diseases of misery, backwardness, and tyranny are so widespread that they seem part of the natural order of things. Hence, there is little to stand in the way of the Soviet Union's identifying itself with the Third World and exploiting its discontent with the West to win friendships through extravagant promises. But the character of the southern flank of the alliance is peculiar; there is always something undefined and fluid in this zone of the world where the conflict between East and West crosses with the northern and southern tensions around a strategically multipolar and ideologically diverse sea.[15] This turmoil may limit Soviet opportunities for influence as much or more than it may damage NATO's.

The movable borders of the European southern flank invite invention of a variety of scenarios. Improvisation is the key term since the leading directors—the superpowers—find it difficult, if not impossible, to impose their rules upon the actors in the area. The proofs are evident in an inventory of conflicts: Franco Casadio has counted sixty-four conflicts and thirty-eight political insurrections in that area since World War II.[16]

An imaginary coastal voyage, leaving the Strait of Gibraltar, would encounter crises at almost every port, especially because of friction between Spain, to which the base belongs geographically, and Great Britain, which has possessed it for many centuries. Even if it is not a dangerously heated issue at this moment, Gibraltar is a problem that divides two NATO allies. Terrorism, present in Spain, France, and Italy, assumes different but always destabilizing forms in all parts of the region. It is another problem for NATO, whether it is in the service of revolution or nationalism.

Then there is Malta. Under British control it served NATO as a naval base and was as significant for denying an important island to an enemy as much as it was a vital headquarters for Lord Mountbatten's command. But an independent Malta government, looking for accords with ideologically similar countries, signed treaties of bilateral defense with Libya in 1976 and with Yugoslavia in 1979. In the spring of 1979, as agreed, the British finally departed and the Maltese archipelago was free to develop the relationship with its new friends, some of whom turned out to be undependable. Full of promises, the Libyan dictator was not punctual in fulfilling any of them. Instead, Libya seemed to be bent on exploiting the relationship as aggressively

---

[15]Garruccio, "Mediterraneo," p. 414.

[16]F. A. Casadio, "Les conflits en Méditerranée" (Paper presented at conference on "Arena of Conflict in the Mediterranean," SIOI Rome, 13 May 1978.

as the larger powers. Consequently, when its North African neighbor attempted to extend its territorial waters to 200 miles from its coast and to within 75 miles of Malta, the latter signed a treaty with Italy in September 1980 that provides Italian guarantees of its neutrality.

One might think that the position of NATO's southern flank would be thereby reinforced since the alliance also would have the use of bases in the event of war. It is worth noting, however, that Malta too signed a treaty of neutrality with the Soviet Union that may be more general and less binding but leaves the question of a Western presence uncertain. For the moment, there is equilibrium, but who can guarantee an internal stability for the future? What then does the treaty actually imply for Italy? Does it imply the availability of immediate bases in case of a threat to Malta? Is the treaty itself even credible in light of Italian dependence on trade with Libya? The answers to these questions are far from clear, but, if it has done nothing else, Malta's troubles with Libya have focused attention on the menace from the South and have moved Malta to a pro-Italian foreign policy.

Continuing on the imaginary voyage, other points of tension are encountered along the eastern Adriatic coasts, as in the case of Kossovo which sees an Albanian minority claiming autonomy in Yugoslavia at a time of great economic difficulties for that country. It should be added that Albania itself may undergo far-reaching change after the passing of its longtime dictator, Enver Hoxha.

On the Aegean Sea, the situation is much worse. There two members of NATO, Greece and Turkey, are locked in friction over Cyprus and over claims to air and naval rights on the sea. (These problems are examined in detail by Victor Papacosma; see below, pp. 189–213.) And on the northern part of the Aegean, Romania deviates from the monolithic pattern of the Soviet bloc, while Bulgaria's ambitions in Macedonia conflict not only with Greek territorial claims but also with Soviet Straits policies. Still further to the south ethnic issues are intense. Foremost is the long Arab-Israeli conflict that provides repeated occasions for Soviet involvement with the part of the world stretching from the Mediterranean to the Persian Gulf and the Red Sea into the Arabian Sea. The destruction of Lebanon is only the most recent manifestation of a struggle that has been in effect since the end of World War II.

Israel's military successes have exercised a modicum of stability in the area, even as they helped to reopen old wounds in Lebanon. At the present moment, Israel finds itself sharing a certain community of interests with the most intransigent of its enemies, Syria, as they both occupy large portions of Lebanon. However, Damascus, which seems now to be the closest Soviet client in the Middle East, is not only anti-Israeli but, because it is still preoccupied with the myth of "greater Syria," it also displays a virulent hostility toward Baghdad, as expressed in its support of the Iranian side of the Iraqi-Iranian war.

Leaving the Asian coast for the African one, new tensions along Libya's borders may be observed. In addition to its difficulties with Malta, Italy, and the United States, its relations with Egypt have worsened over Libyan aggressive movements to the south, endangering the sovereignties of Chad, Sudan, Uganda, and Niger. The three Maghreb countries in the western zone of the Mediterranean seem more stable, both in their domestic and foreign relationships, but, if Tunisia and Morocco are clearly associated with the European Economic Community (EEC), it is necessary to underline that the policies of Algeria are fully committed to the Third World and open to Soviet manipulation. Moreover, the relationship between Algeria and Morocco has been disturbed by Saharan territorial problems and by the support given to Moroccan rebels by Algeria during the Polisario war. Finally, it should not be forgotten that Morocco also has its "Gibraltar" under the names of Ceuta and Melilla; the antagonist in this instance is Spain.

There is an additional element of potential trouble between the central and western zones of the Mediterranean where the United States has access to the bases of Morocco and the Soviet Union to the bases in Libya. Although access to the Moroccan bases for the Americans is stronger than the Libyan commitment to the Soviets, it is necessary to note that Morocco is armed by the United States and France but has developed important economic relationships with the Soviet Union. Algeria is mainly armed by the USSR, but it trades with the Americans and, above all, with the French, with whom it is linked via gas pipelines.[17] These are far from uncomplicated relationships.

From all this, it is clear that the Mediterranean Sea is assuming an increasing value for Western Europe. Mediterranean politics on the part of the EEC is, at the same time, confirmation of a relationship between Western Europe and the Mediterranean which should become pivotal. Although the importance of the coastal countries of the east and south may seem limited in the economic and energy fields, the security and the stability of all the basin remain fundamental for the freedom of the Continent. The idea, spread in some European circles that it is possible to maintain a favorable status quo vis-à-vis the Soviets, should not induce complacency; the area is too important to the survival of the West to permit relaxation of attention. In the past the balance of power has been changed by external events, as witness the Iranian and Afghanian episodes. Without question, tension is greater in the Persian Gulf area today than in any part of Central Europe.

From a general viewpoint, an attack on the Gulf is an indirect attack on the center of Europe. There is a direct and very strong link between the defense of the Mediterranean and the security of the European countries of the alliance. The southern flank has become preeminent for NATO, although the maintenance of the present perimeter of the alliance is insufficient to secure free access to those resources of the Middle East vital for the industrialized countries of Western Europe.

---

[17]E. Osswald and W. Wessels, "European Concepts for the Mediterranean Region"; W. Zartmann, "Maghrebi Conflicts and Mediterranean Implications" (Papers presented at IAI Castelgandolfo conference, 6–9 September 1982).

CONCLUSION

The objective of the reinforcement of the southern flank must be that of rendering the alliance strong enough to defend the Mediterranean and to maintain stability on its flanks.[18] The Mediterranean today has become a smaller sea in geostrategic terms as a consequence of the rise of air power and availability of new technology. They have made the fleets more vulnerable and decreased the efficiency, of which on many occasions in the past had inhibited conflict.[19] Since the Soviets consider military power, on the whole, as the most important instrument of foreign policy, it is necessary that they receive credible messages in such a way that they do not commit mistakes.[20] The Soviet bloc must not be misled by the diversified approach that the main Western European nations take toward the Mediterranean countries.

NATO remains the structure best equipped to cope with Soviet pressure but only if its message is clear and its actions are coordinated with other bodies, such as the European Community, that have their own complementary ties with the Mediterranean. Still, it must be observed that such reinforcement can only be complementary to NATO's actions, not substitutes for them. While Europeans and Americans may differ over methods of approaching the southern sector's problems, their differences should be resolved within the alliance rather than in the European Community where economic issues distract attention from security matters.

The key to peace and security in the Mediterranean is the appropriate use of the special geostrategic positions presented by the member nations. A glimpse at the entire situation of the southern flank seems to confirm that. In the western basin, the fundamental strategic role of the Iberian peninsula, linked with the American bases of the Azores and Morocco, serves the continental French-Italian coasts. It does not seem to be an exaggeration to speak of a Western predominance in case of emergency in that geographical sector. Taking Italy as the hinge of the Mediterranean, it is easy to see that of the two maritime passages eastward, the Straits of Messina place the northern route squarely under NATO control, while the southern route—the Strait of Sicily—is shared between Italy and a friendly, if neutral, North African nation, Tunisia. Passing beyond the hinge, there are no enemy bases against NATO, aside from Libya, which could be attacked from the sea and, according to the circumstances, also from the land. Moreover, it is far from certain that Libya wants to engage NATO in a Mediterranean conflict.

---

[18]Ibid. See also C. Nigoul, "La sécurité en Méditerranée," *Défense Nationale* 37, no. 6 (May 1981): 61–64; and C. E. Zoppo, "American Foreign Policy, NATO in the Mediterranean, and the Defence of the Gulf" (Paper presented at IAI Castelgandolfo conference, Croom Helm, London, 1984).

[19]M. Cremasco, "The Military Presence of the Riparian Countries," IAI Castelgandolfo conference.

[20]C. E. Zoppo, "The Changing U.S.-Soviet Military Balance and Security and Politics in the Mediterranean," Toulouse conference. For further reference see T. C. Schelling, *Arms and Influence* (New Haven, CT: Yale University Press, 1966).

The problems of the alliance are more difficult in the eastern basin where a relatively thin segment of Greek-Turkish land covers the Aegean Sea from a northern direction. Not only is it closer to the Soviet Union, but it is also wracked by internecine trouble that invites Soviet intervention. Yet this problem of the southeastern NATO perimeter presents an interesting example of that earlier mentioned combined strategy. The use of the U.S. Sixth Fleet—a dominant military power in the Mediterranean in its own right—has changed. Originally its major purpose was to sustain Greece and Turkey in the eastern Mediterranean; it is now employed beyond Suez as well. Such an operation is made possible by the collaboration of the other Western powers. This is just what the Italian, French, and Spanish armed forces have undertaken in making new commitments in the area. The presence of Italian, French, and English soldiers in Lebanon, Italian naval units in the northern extremity of the Red Sea, and British and French naval forces in the tense southern areas is dramatic proof of this new stance. It is now conceivable that in the common interest a multinational Atlantic force in the Mediterranean may operate beyond the responsibilities of the single countries. Together the NATO partners could inhibit Syrian aggression in the eastern end of the Mediterranean and Libyan aggression on the southern flank.

In the military field the increasing importance of the air bases, along with the availability of the "unsinking aircraft carriers"—the islands and southern peninsulas of the European continent—favor a Mackinder rather than a Mahan approach to regional security. The role of sea power is diminishing; political and economic instruments also must be used. Military stability, even if it were possible, is not enough. The West must mobilize all its resources to cope with problems of emerging nations as well as with Soviet ambitions.[21] Such an example can be found in the new Italian policy of economic cooperation in which Africa claims first priority. Within limits, it offers a positive approach to the Third World. No less positive would be the application of Article 2 of the North Atlantic Treaty to the economic growth of the emerging countries in adjacent areas of the Mediterranean. Its emphasis upon encouraging economic growth within the Atlantic Community could be extended to areas within and beyond the Mediterranean. The result not only would be a strengthened alliance but also a more secure and prosperous southern flank.

---

[21]See Garruccio, "Mediterraneo," pp. 422–25; F. A. Casadio and A. Mura, "Lo strumento militare Italiano e la mutata situazione politico-strategica nel Mediterraneo" (Paper presented at LUISS Permanent Seminar of Strategic Studies, Rome, July 1981); Nigoul, "La sécurité"; Labrousse, "Sécurité en Mediterranée"; Cremasco, "The Military Presence"; *NATO Today: The Alliance in Evolution: A Report to the Committee of Foreign Relations U.S. Senate* (Washington, DC: Government Printing Office, 1982); P. Olsen, "Some Considerations in Designing a Strategy for NATO Southeastern Flank" (Paper presented at conference on "The Southern Flank of NATO, the Mediterranean and the Persian Gulf," Naples, 21–23 September 1981); Aliboni, "Il fianco"; P. H. Mercillon, "L'armement est-il un facteur d'équilibre ou de déséquilibre dans le bassin méditerranéen?" (Paper presented at Toulouse conference 1–4 October 1980); and M. Carnovale, "Interdipendenza economica e sicurezza nel Mediterraneo," *CSST-Bollettino di informazione* 5, no. 6 (November 1982): 9–10.

THE IBERIAN CONNECTION

Portugal's Special Relationship: The Azores, the British Connection, and NATO

ALBANO NOGUEIRA

As far back as the Middle Ages, Portugal and England had concluded a treaty of alliance. However, Winston Churchill's statement, made before the House of Commons on 12 October 1943, may help bring the relationship between the two countries into a sharper perspective:

> I have an announcement to make to the House arising out of the treaty signed between this country and Portugal in the year 1373 between His Majesty King Edward III and King Ferdinand and Queen Eleanor of Portugal. I spoke in a level voice, and made a pause to allow the House to take in the date, 1373. As this soaked in, there was something like a gasp. I do not suppose any such continuity of relations between two Powers has ever been, or will ever be, set forth in the ordinary day-to-day work of British diplomacy. . . . This treaty . . . was reinforced in various forms by treaties of 1386, 1643, 1654, 1660, 1661, 1703, and 1815, and in a secret declaration of 1899. In more modern times the validity of the Old Treaties was recognized in the Treaties of Arbitration concluded with Portugal in 1904 and 1914. . . . This engagement has lasted now for nearly six hundred years, and is without parallel in world history.[1]

In spite of such a procession of treaties, it would be a futile exercise to attempt to sort out a clear-cut definition of the inner element of continuous accommodation to common interests that made possible successive adjustments to the demands of new circumstances. It is clear, however, that the community of interests, founded on geopolitical affinities, has been the basis of Anglo-Portuguese relations in modern times. In May 1944 Prime Minister Oliveira Salazar characterized it succinctly:

> Let us leave aside the lack of precision of the alliance, . . . and let us look only at its line, which is due more to its centuries old existence than to its letter on texts laying down in the archives. Reduced to what we may consider its essence, the alliance is little more than the mutual

[1]Winston Churchill, *War Memoirs*, vol. 5, *Closing the Ring: The Second World War* (Boston: Houghton Mifflin Company, Boston, 1951), p. 165.

> and permanent understanding of a certain peninsular and colonial position, to us, and of a certain amount of services guaranteed thereby, to England.[2]

SIGNIFICANCE OF THE ALLIANCE WITH GREAT BRITAIN

It is not the purpose here to place the alliance in a historical perspective, nor is that necessary. In the not so distant past, and in the relatively narrow span of slightly more than two decades, it was put to the test of a number of events and situations rich enough in variety and significance to make the correct appraisal of its vitality and resilience a relatively easy task. When confronted with friction and occasional dissension, the alliance met the challenge.

During that period Portugal was ruled by a single man, António de Oliveira Salazar, prime minister (1932–68) and foreign minister (1934–47). His indisputable ability and long tenure were largely responsible for the consistency and continuity of Portugal's foreign policy. Fortunately for historians Salazar's scholarly habits, acquired from his academic background, produced frequent appraisals, both of the state of the world and of Portugal's role in it. It is worth noting that the British alliance appeared frequently in his public statements and that at no time in the difficult years between the outbreak of the Spanish civil war through World War II to the signing of the North Atlantic Treaty did Portugal seek to change the nature of those ties.

Among prewar historians and scholars the opinion seemed to prevail that most Portuguese involvement in European conflicts had not proved beneficial to the country. Salazar held the same view and in 1935 he wrote:

> . . . the Portuguese nation, whenever it intervened in quarrels or fights outside the Peninsula, did so, so to say, accidentally and seldom in defense of a vital interest. We are above all an Atlantic power, tied up by Nature to Spain, politically and economically leaning on the sea and the colonies, ancient discoveries and conquests.[3]

That idea became a *leitmotiv* in Salazar's public statements and one of the fundamentals of the truly independent foreign policy he consistently sustained: that Portugal, an Atlanticist rather than a Europeanist was, nevertheless, a European nation.

---

[2]Oliveira Salazar, *Discursos e notas politicas,* 6 vols. (Coimbra: Coimbra Editora, 1951), 4:62.

[3]Ibid., 2:79–80.

> Under these conditions there is a feeling that the traditional line of our foreign policy, coincident with the true interests of the Portuguese motherland, lies in our non-involvement, when possible, in European disorders, in maintaining the peninsular friendship, in developing the potentialities of our Atlantic power.
>
> . . . from the same character of an Atlantic power the British alliance was also born, an alliance which common interests have maintained through the passing of centuries, undamaged by the vicissitudes (normal) in those kinds of ties.[4]

When Salazar made these statements in 1935, he could not have anticipated new vicissitudes in the form of the Spanish civil war which would strain the alliance, but he certainly had in mind the effect that the recently installed republic in Spain could make in the relations of European democracies with the Portuguese regime. The community of interests covered by the alliance being duly acknowledged, the proviso, nevertheless, was appended that "our actions both in internal and foreign policies" were not to be hindered. No confusion should exist. By the same token, divergent interests could coexist beyond those covered by the alliance. No doubt the old bonds were severely strained by Portugal's vigorous support of Franco's Spain. In Britain political forces opposed to the Portuguese regime tried to take advantage of the situation. With characteristic disdain, Salazar dismissed those who

> from time to time in England . . . [those] without great responsibilities to be sure, angry about (Portuguese) attitudes in other fields, are putting forward, more frequently of late, above the great national and international interests, their passions and resentment and invite the English Government to review the problem of the alliance with Portugal.[5]

While once again stressing the "secular friendship and alliance," which he called "an element of foreign policy of the highest value," Salazar reminded the British of the words of Lord Palmerston, to whom the advantages of the alliance with Britain were "many, great and obvious," and "commercial, political, military and naval." If lost, "some of them would not be merely lost"; they could become "formidable attack weapons against us in the hands of a hostile power." No matter how greatly conditions may have changed, he observed that Great Britain was obliged "to keep open the South Atlantic communications, the Mediterranean and the Cape route to the East." In that, Portugal has a role to play: "In the mainland, in the Islands and in the Colonies, she is at the confluence, and on the ports of call and security, of the great maritime routes." In short, "it is in fact as an exchange of utilities and services, which we persist in believing to be vital to both nations, that the alliance had

---

[4]Ibid., pp. 80–81.
[5]Ibid., p. 307.

worked without any hindrance or attritions other than those originating in little family squirms."[6]

Exactly why the alliance remained valid, alive, and effective throughout the centuries is due to several factors. First, it owed its existence to Portugal as "an Atlantic power"; second, Great Britain possessed similar characteristics in its insularity and overseas projection to the empire; third, the lack of precision strengthened the usefulness of the alliance; fourth, the alliance took the form of an exchange of utilities and services; and, finally, it did not prevent the two parties from holding other friendships. This last point is of particular importance since the alliance had much to do with Spain's signature, both on the 17 March 1939 treaty of friendship and nonaggression and on the 29 July 1940 additional protocol, as well as with the position Portugal would be expected to assume should war break out, and the position it actually assumed at the outbreak of any conflict.

The treaty with Spain was born under a shadow of suspicion, prejudice, and apprehension; "Red" Madrid had yet to surrender to Franco but did so ten days later. Three days before ratification Spain joined the Axis powers in the Anti-Comintern Pact. No doubt the treaty was an offspring of Franco's impending victory, which in turn led the world to regard it as a mutual undertaking whereby the two parties would help each other preserve their authoritarian regimes from external threats. However, the treaty far transcended the issue of mutual protection; in fact, the idea for the treaty came from Franco. On 16 September 1938 his envoy in Portugal was, at his request, received by the prime minister "to express . . . concern with a possible conflict in Europe and the Portuguese attitude in such an [event]." He gave Salazar the assurance "that Spain was not prevented from keeping her neutrality by any engagement with Germany or Italy. . . . Assuming that the intention and convenience of Portugal would be likewise," he would suggest, "on behalf of his Government, a pact of non-aggression between the two countries."[7]

All things considered, had not such an understanding been formed between the two countries, they might have found themselves fighting on opposite sides. The treaty thus may have played its part in containing the war beyond the Pyrenees to the clear advantage, eventually, of the Allied powers. Reactions abroad were complimentary: In London Halifax said that the treaty "contributes to the reestablishment of conditions of peace and stability in the Iberian Peninsula"; in Belgium it was believed that the pact was "of utmost importance to the clarification of the Spanish case"; and in Washington the State Department expressed satisfaction. Germany and Italy congratulated Franco.[8]

In the preamble of the treaty the two countries acknowledged the "community of feelings" and their mutual interests, "stemming from geographic

---

[6]Ibid., pp. 307–08.
[7]Franco Nogueira, *Salazar,* 6 vols. (Coimbra: Atlantida Editora, 1978), 3:175–76.
[8]Ibid., p. 192.

and historic realities," and expressed their conviction that "any contribution to the maintenance and assurance of peace between [them] . . . would constitute an important factor to European peace." They stated that any obligations so far assumed by them "in relation to third powers" were neither "contrary to the development and strengthening of their mutual relations," nor did they "contradict the designs and clauses of the Treaty." Therefore, they bound themselves, "to the absolute respect for their frontiers and territories," not to indulge in any act of aggression to or invasion of the other, not to give aid or assistance to an eventual aggressor or to allow an aggression or attack against the other to be launched from an ally's territory, not to enter into a pact or alliance against the other, and to safeguard these commitments in any pacts or alliances that might be signed in the future.

Not until two months later on 22 May did Salazar comment on the treaty and then only in the wider context of an appraisal of his country's position in the international scheme. He emphasized that Portugal is an Atlantic country "as only England can be." So, like England, Portugal holds great interests and has responsibilities to perform on other continents and seas. What, therefore, could Portugal claim from Europe, or to what extent should the Portuguese consider themselves part of European life and culture? Salazar's answer was blurred in deliberate vagueness but full of far-reaching innuendoes: "We are politically linked to Europe insofar as we may be pushed by the unforeseeable consequences of a generalized conflict, and morally insofar as it may continue to be the brain and the heart of the world."[9]

What Salazar said of the treaty was clearly meant to be heard and understood by audiences outside the frontiers of the country. His assurances were pointed toward those who were in doubt: Spain "had succeeded in killing in her own blood the virus that had threatened the peace and civilization of the Peninsula; . . . in the name [of the principles of her social and political order] she can affirm that, having rebelled against communist servitude, she had not fought so gallantly to otherwise mortgage her own independence and destiny." Obviously, Salazar intended to reassure those abroad who were fearful of a new Spain dependent in the future on Germany:

> There is only one limit today to the full liberty of Spain in her external action—the Treaty of friendship with Portugal. And, if, to us, that is in a way highly gratifying, on the other hand the fact that the definition of any policy by Spain had been preceded and conditioned, in the minds of her highest leaders, by the declarations in the Portuguese-Spanish pact, shows a perfect understanding of the superior interests in the Peninsula.

This last comment was pointedly accompanied by a message to Great Britain:

---

[9]Ibid., p. 139.

> Those who may have reflected on the traditional English policy and on the essentially defensive sense of its international conduct would be prepared to understand how much England should welcome the setting up of this true zone of peace in the Peninsula, given the fact that one of the States is her own ally and the other had always been her friend.[10]

Not sure of what the future would reserve for Britain and France after Munich, Portugal presented a circumstantial argument: a Red victory in Spain would have put English and French interests on the peninsula at constant risk of collision because of the ideological and political conflict on the one hand and the Anglo-Portuguese alliance on the other. In fact, "only with a Nationalist Spain, brotherly linked to Portugal, may France and England work for the security of their interests and frontiers and for the betterment of their relations. In this light, our Treaty with Spain is the crowning to a work and the cornerstone of a policy." Then he clearly expressed that Portugal intended "to maintain herself within the constants of her history, assuring in fidelity to the Anglo-Portuguese alliance, the defense of common interests and safeguarding, in what is extraneous to it not only liberty of action but also the existence of many other friendships."[11] Salazar did not hesitate, nor did he mince his words, in vouching for the Spanish unwillingness to take part in the impending war. By the same token, he tried to show Britain that, by signing the agreement with Spain, Portugal was indirectly rendering a service to its old ally.

An additional protocol to the treaty was signed a few weeks after the surrender of France, and, because of that disturbing event, the two Iberian countries came forward to express their intent to give "permanent increasing efficacity to the Treaty." In conformity with its spirit, they would "undertake to concert . . . on the best ways of safeguarding as much as possible their mutual interests, whenever facts be foreseen or occur which by their nature could put in jeopardy the inviolability of their respective territories on the Continent or be a danger to the security or independence of one or the other Party." That could be done on the initiative of either party "whenever a fact [of such a nature should] occur or is considered possible." A new probable situation was therefore contemplated: the treaty only provided for decisions taken by the parties, while the protocol covered situations derived from actions taken by external forces. With the sudden fall of France, little of consequence was known of further German plans in the war against Great Britain, and it would not be alarmist to expect the worst.

The news of the Munich agreement had pleased Salazar, who was particularly effusive in his congratulatory message to the British. The outcome of the crisis had confirmed his conviction of the "improbability" of war and relieved him of the fear of seeing it spread. He had not understood why

---

[10]Ibid., pp. 149–50.

[11]Both citations are from ibid., pp. 146–51.

Portugal should have been involved in a general conflict but, at the same time, was glad that his country had been spared from such "horrors of misery and destruction." Although presumably a mere reflection of an event that could have happened but did not, Salazar's words were meant to be part of an indirect method of making known Portugal's decision to avoid any involvement in a European conflict. This position was upheld in 1939 and led Salazar to be deliberately vague in the way in which he made public Portuguese neutrality when the war broke out. In fact, he did not abide then by the international usage of proclaiming it. What emerged was simply a matter-of-fact statement that his country would remain in a situation of neutrality; in other words the status quo would continue. At the same time, it was not in any way a disavowal of the British alliance nor of the obligations thereof. That is why it was acknowledged that the status quo had not been affected so far. It was thus the expression of an expectant attitude, subject as much to a change or reappraisal should compelling circumstances arise. According to Salazar,

> . . . some of the greatest nations of our Continent—friends of ours some of them and one our ally—will be at war. . . . Fortunately the duties of our alliance with England, from the confirmation of which we do not want to excuse ourselves . . . do not force us in this emergency to abandon the *situation of neutrality*. . . . The Government hopes that the interests of the country, or its dignity, or its obligations, will not constrain it to compromise this situation.[12]

The pledge, in which deference to the old ally was coupled by the recognition of the permanence of the alliance, came to be honored first in 1943 when Great Britain actually asked for much-needed facilities in the Azores in order to reduce the level of damages inflicted on Allied merchant shipping by German submarines operating in the zone, and then in 1944 when, on the ally's request, exports of wolfram were prohibited to all destinations. This and the fact that in 1944 facilities in another of the Azores Islands were granted to the United States led to the coinage of "collaborating neutrality," an interpretation of the Portuguese attitude toward the war, an attitude with which Salazar was not displeased.

THE SHAPE OF THE WORLD TO COME

In May 1944 the Portuguese prime minister went before the National Union, the only political organization allowed in the country and the sole supplier of deputies to the National Assembly, to speak at some length on

[12]Ibid., p. 174.

the preparation of the nation for the postwar world. This was a tantalizing subject since the vast majority of his audience was strongly in favor of the Axis powers. Significantly, Salazar approached the subject under the hypothesis of an "unequivocal German defeat." To him, prevailing conditions seemed to be leading to the emergence of three great hegemonies, at least in the European, Asian, and American areas: those of the British Empire, the United States, and the Soviet Union.

"The unfolding of events," Salazar said, "seems to lead to the destruction of a concept of Europe" which was wrongly "built on a pure Continental basis" and was nothing but "an amputation of its moral being, . . . contrary to history and to the grandeur and civilizing power . . . of the Continent . . . . a perennial source of civilization and a generous begetter of nations." On the other hand, it also appeared that "Europe was irremediably dividing itself, and impoverished, losing positions and weakening by the sole fact of the war. The problem now is that it must not continue carelessly wasting the conditions within its grasp for the sake of Western life. Fortunately, the whole of Africa is a Western European dependence, the two facing America from pole to pole, making the material basis for the role it ought to continue to play in the World." From this, the prime minister extracted a not too pessimistic conclusion:

> The circumstances are evolving in such a way that one of the greatest centers of World politics, particularly insofar as the United States would consider it their interest and their duty to help Europe to raise from the ruins of war, will be placed by the force of things on the vast Atlantic, and nothing but for that the coastal countries will be called to play a predominant role: England, France, the Iberian Peninsula, the United States, South America . . . will be called to an intense collaboration, and, through it, Western Europe to be one of the fulcra of orientation of politics in general.[13]

Besides a perceptive, if not infallible, vision of a new world order, Salazar proposed some ideas concerning his country's role. Among them were closer ties with Western Europe, a new position as fulcrum in a Euro-American system centering in the Atlantic, a revaluation of specific Portuguese interests in the colonial territories, and a continuation of the special relationship with Britain. In such a context, the prime minister could not help but raise the question of whether or not there were still good reasons for the alliance to be maintained with Great Britain. The answer was, once again, very vague indeed: if it served Portugal, it served England, and because it served England, it served Europe.

In spite of the lack of precision, a subtle concession appeared in Salazar's references to Western Europe, but he added that "the adjustment [of the

---

[13]Salazar, *Discursos e notas politicas,* 4:60–61.

alliance] to the general orientation [he had outlined] is so fitting and perfect that there was no need of retouching by whatever form of organization [to be assumed by] the international society, should this be maintained within the realities of life and not lost in a dreamland."[14] A few weeks later the British foreign minister told the House of Commons that the Portuguese "alliance had been for a long time highly beneficial to both Parties." The wish of Her Majesty's government was that "these unparalleled relations, ever increasing through the passing of the years, remain a link between the Government and the peoples of the British Commonwealth and Portugal, as strong in the future as they have been in the past."[15]

No doubt the question of the survival of the alliance had some internal undertones for Salazar. His opponents, with help from the opposition in Britain, were eagerly expecting that the victory of the democracies over the dictatorships also would bring about the downfall of his regime. At any rate, he made it clear that, regardless of how the international organization was to be structured, Portugal had to assume that certain factors were likely to confer a new international importance and heavier responsibilities on the country.

At the end of war in Europe, an exhilarated Salazar made a speech in the National Assembly, congratulating the Allies on their victory and expressing his particular satisfaction that Britain was at the forefront of the victorious nations. To the stupefaction and incredulity of many, he then vented some hitherto repressed feelings: "I humbly confess that my doubts were only converted into a certainty when observing that their war effort, while within the reach of the extraordinary possibilities of the British people, raises a doubt that it has ever been achieved in the History of mankind." Once again the variation on an old theme is brought forward, but this time Salazar was more emphatic in pointing out that to "a perhaps much too continental vision of Europe" is being superposed "an historically more exact conception of its universality. . . . The victory of England and America will move the center of gravity of international politics into the Atlantic with a resulting impact on Western thinking."[16]

One week later, and again before the National Assembly, Salazar presented an inventory of the past, claiming that it was the end of an era in history and underlined two points. First, Spain had been a valuable shield for Portugal's own neutrality, as much as Portugal had been for Spain's; and second, it was time to replace Portuguese independent neutrality with membership in a strong alliance. It seemed to him that World War II would be the last time it could and should be neutral in a European conflagration.[17]

The year 1945 brought other satisfactions to the prime minister. The recognition of the "nation" as a concept basic to the new world organization—

---

[14]Ibid., p. 62.

[15]*Times* (London), 8 June 1984.

[16]Salazar, *Discuros e notas politicas,* 4:97, 103.

[17]Ibid., pp. 104, 107–08.

the United Nations—pleased him enormously. As jurist and leader of a small "nation," his concern with the preservation of the sacred "national principle" was never concealed. In his own words, the universal reiteration of the "existence of nations, different from one another, independent and free, and organized in States, sovereign and equal," was not incompatible with an eventual adoption of an "aristocratic principle for the effective direction of society."[18] The point he was making was that the actions of the great powers should be mitigated by those of the smaller ones.

Further on he praised the admission of regional agreements by the new organization: "Prevailing realities recommend them; they safeguard for [Portugal], as a far reaching instrument [under such a UN sanctioned system], the British alliance; they allow that the already close relations with the United States, France and the colonial neighbors . . . be further developed." Again, Salazar called attention to the shift in the center of gravity of international politics to the Atlantic, but now he emphasized that "deeply rooted here and in Africa, on large Atlantic coasts . . . our place is well established."[19]

In the immediate postwar years, Portugal did not actively intervene in world affairs. Its master and ruler assumed the position of spectator; he was an attentive observer of the settlement between victors and the defeated rather than an actor with any contribution to make. Salazar soon came to the unpleasant conclusion that the hegemonic triad he previously had visualized was losing ground and being replaced by a bipolar world system, with the United States and the Soviet Union at the top. Britain, he assessed, had come out of the war in a changed position relative to the others. It was certainly still able to act independently in defense of its own interests, but otherwise it had become a part of the dual Anglo-Saxon subsystem. The higher ranked United States, unlike in 1919, felt the responsibility of its force and victory and was ascending to the highest role of world political leadership, by their own worth no doubt, but also because they were forced into it by their friends. It was, he believed, almost a hegemony by plebiscite.

While the prime minister regarded the United States and Britain as pacific countries, Salazar doubted if the same could be said of the Soviet Union. No one could deny that the USSR held strong power and was endowed with capabilities of decision and action with which no other country could compete. It also was free from the weight of public opinion which, in the Soviet Union, was nonexistent or easily manipulated to support the regime's policies. Furthermore, the USSR through its satellite governments "holds now every possibility to dominate Europe entirely and is aware that she can do it without meeting with any resistance."[20] All this became afterwards a new *leitmotiv* of the prime minister's world outlook: the unfolding of events was confirming him both in his views and his concern. By November 1947, Salazar

[18]Ibid., p. 113.
[19]Ibid., pp. 113, 251–53.
[20]Ibid., pp. 289–91.

could use the Red menace to justify his wartime stance and indeed to blame the West for postwar problems.

> Within a few years no one will deny how erroneous the Anglo-Saxon war policies had been. All will come to agree that the errors committed stemmed from two others: the thesis of unconditional surrender and the priority given to the European theatre of operations. Germany had been smashed to such an extent that she will be in no position to play her part as a factor for equilibrium in Europe. Worst of all, Russian advance to, and occupation of, the heart of Europe and its dominant positions became unavoidable. . . . Under these terms Europe is condemned to live for years on the alert and in a state of unrest.[21]

As the political chasm was steadily widening in Europe, Salazar was trying to portray himself as an interpreter of its own uneasy conscience. On 28 April 1948 he reiterated his warning of impending dangers and the urgent need to confront them. Speaking of the West in face of Russia, he said:

> At the end of the war, a great and powerful nation went on increasing and consolidating its strength, and came to assert, by the presence or threat of that strength, a line of thought, which at one time might have been considered preventive and a reinforcement of its security, but, beyond that, cannot but be conceived as an imperialistic tendency, clearly hegemonic. I refer to Russia. . . . The other Allied powers were caught by surprise . . . and seemed to rely on victory as if victory would mean peace. The former went on, developing its strength and utilizing its positions, as if the war were still to proceed. . . . To this no other explanation exists but the following: the Western Powers considered their war goals attained with the smashing of Germany; not so for Russia, which, farther than the reinforcement of her defense and the increase of her strength, looks forward to the dream of worldwide revolution.

To this he appended a corollary: "From [all of this] I extract two illustrations: first the West will not leap deliberately into a war against Russia; second, the West ought to use its utmost efforts to restrain Russia from leaping into a war against it."

Less than one year later, on 7 January 1949, Salazar anticipated, in certain respects, the idea of the Atlantic Alliance when he commented on the inadequacy of the United Nations in dealing with the outstanding problems of the world. Accepting what he believed to be a Western consensus, he expressed his agreement with "the necessity of, without openly going against the [U.N.] organization, to start disassociating fundamentally political problems, such as those of defense and economic rehabilitation, in order to cope

---

[21]Ibid., pp. 327–31.

with them on bases either regional or of closer affinities of interests or civilization, rather than on any other basis provided by the universal Areopagus."[22] With the Truman Doctrine and the Marshall Plan less than two years old, the North Atlantic Treaty was to come three months later. In a way, the acceptance of the treaty by Portugal was some proof of a theorem that the prime minister had been obstinately developing for a long time.

WHY PORTUGAL JOINED THE TREATY

Accession of Portugal to the treaty appears not to have been placed in serious doubt because of its nondemocratic authoritarian, or dictatorial, regime. President Harry Truman alluded to the question of membership in his memoirs, noting that "enemy occupation of the territories of Norway, Denmark, Iceland, Ireland and Portugal (with their dependent territories in the Atlantic area, such as Greenland and the Azores) would represent a threat to the security of Western Europe."[23] He was so casual about listing Portugal that it was obvious that the problem of neutrality in World War II, or of having an authoritarian government, would not be barriers to membership in the alliance. In fact, the case of Portugal was to be pragmatically handled. Irrespective of any judgment that may be passed on the Salazar dictatorship, the point was that the Allies recognized that the participation of Portugal would not only protect the Azores more effectively from the hazards of an enemy occupation but also conveniently would assure their outright inclusion in the system. Could the contribution, given to the victorious outcome of World War II by the facilities granted to the Allies, be ignored? Because this subject has been written about in detail elsewhere,[24] it suffices here to reiterate that the invitation to Portugal and its acceptance showed a commendable statesmanlike spirit on both sides. Each made its selective reading of the principles set forth in the preamble and met on the common ground of the high purposes to be pursued.

When the prime minister went before the National Assembly on 25 July 1949 to ask for authorization to ratify the treaty, Salazar elaborated extensively on the goals sought but not reached by the Allied powers in the war and the wrongs done by compelling Germany to surrender unconditionally. The war, he recalled, had been fought to prevent the installation of a German hegemony in the center of Europe. However, by demanding nothing short of total surrender, the Allies had caused Germany to be dragged into total disintegration

---

[22]Ibid., p. 376.

[23]Harry S. Truman, *Memoirs: Years of Trial and Hope, 1946–1952* (New York: Doubleday, 1956), p. 248.

[24]See Albano Nogueira, "The Making of the Alliance: A Portuguese Perspective," *NATO Review* 28 (October 1980): 10–14.

and nothing had been attempted immediately to fill the gap. The Soviet Union thus had been given the opportunity to install its hegemony in much of the area, all this when most of the Western nations were in a state of economic distress, moral fatigue, and internal disintegration. The United States, Great Britain, and France had been able to impose their will on the power against which they fought, but then they could do no more than to preserve the Western zones of Germany from Soviet interference. This was not so on the Soviet Union side which was strong enough to dictate in the north, east, and southeast of Europe. The Soviets seemed quite prepared to bring their armies to the heart of Europe and could move virtually unopposed to the English Channel and the Pyrenees. They seemed equally capable of attaining a kind of world domination. With a bit of sarcasm, Salazar commented: "Glory had shined to some; victory has befallen to others."[25]

In such a state of affairs, the prime minister, although he paid due tribute to the insufficient efforts previously made by Western European nations to organize themselves against the Soviet Union, stressed the importance of American and Canadian policy which brought to Europe the indispensable force for an effective defense of the Continent. Salazar praised mainly the American attitude for showing a strong feeling of solidarity and assuming their responsibilities as political leaders of the world.

As noted, it was a recurrent idea in Salazar's reflections on the foreign policy of his country that there was a dichotomy between Portugal's physical existence on European soil and some sort of Atlantic destiny. This duality historically had led to a disinclination to become enmeshed in the Continent's troubles. This trait had been a constant factor in the alliance with England; it was a persistent and latent noncontinental affinity binding the two countries. In such a context, the North Atlantic Treaty would insert Portugal into a new framework. He believed it to be his duty to clarify whether or not entry of Portugal into the treaty organization should be considered a deviation from traditional policy. The alliance, after all, had been inspired, he argued, by the necessity of protecting the maritime routes fundamental to England and by an essential element of Portuguese life on the Continent as well as in the other parts of Portuguese territory scattered around the world. Fundamentally, Salazar asserted, the alliance with Britain had always functioned as a promise of mutual support aimed at maintaining the freedom and security of the Atlantic.

> When this line was not followed, . . . leading to Portuguese intervention in internal conflicts in Europe, . . . contrary to our interests and without any profit, we came back home . . . always deceived by our intervention. . . . These deviations from our guiding principles as defined by the vital needs of the Portuguese people—facing the sea, back to the

---

[25]Salazar, *Discuros e notas politicas*, 4:405.

> land—cannot be taken as precedents to be followed, but rather to avoid in so far as it would be possible.[26]

The key to all of these problems lay in the involvement of the United States in Europe and that, in turn, rested upon the American assessment that the defense of the European coastal countries was in its own self-interest. Since neither internecine conflict within the alliance nor a rebirth of German nationalism were likely in the future, the one potential threat to stability was an attack from the East. Such potential dangers to Portuguese sovereignty or freedom of action were satisfied by the special understandings that accompanied the signing of the treaty.

In compliance with Article VIII of the treaty, the Portuguese government declared in Washington that the treaty with Spain was compatible in principle with the commitments now assumed. The point was extensively discussed with the Madrid government, which had been uncomfortable with Portugal's adherence to the treaty. Spain's sensibilities were soothed by Salazar's championing of Franco's cause in the alliance and by professions of NATO's compatibility in principle with the Iberian Pact, although openly admitting that commitments assumed under the treaty had to be viewed in the light of the principles set forth in the peninsular agreements.

All this gave the prime minister the opportunity to deplore Spanish exclusion from NATO. His criticism was based on three points. First, the absence of Spain left a geographic and strategic gap; second, it had not been sufficiently taken into account that Spain in fact would be able to render an authentically important contribution; and third, the practical value of Portuguese membership would be different were Spain in or out of the treaty. To this Salazar added: "Full action by a Western front against a possibility of aggression is highly conditioned by an identically oriented policy in the Iberian Peninsula. By luck or enlightened purpose, the Treaty is flexible enough to adapt itself to more than a situation." The "unsurmountable" realities that excluded Spain from the alliance made it more compelling for Portugal to insist on, even at the risk of being confronted with the incomprehension of others, advocating not only an invitation to Spain "to cooperate in the various international organizations dealing with European problems" but also to assert repeatedly that Spain should be invited to join the treaty. Should political difficulties continue to block it, some other understanding should be found.[27]

Years later at the United Nations, after the admission of Portugal in the famous "package deal" of December 1955, Portuguese "colonialism" for a long time was discussed, dissected, attacked, and condemned. As a result, Portugal experienced comparable difficulties within the NATO community which led to cool relations with fellow members of the alliance and to limits

---

[26]Ibid., p. 413.
[27]Ibid., p. 417.

on grants of military aid. Yet the realities of the long-standing British connection and the significance of the Portuguese Atlantic islands always prevailed over misgivings about the Salazar regime's ideology and its particular African colonial policies. Portugal was a member in good standing at the time of the prime minister's death in 1970.

A NEW PORTUGAL

On 25 April 1974, Professor Marcello Caetano, Salazar's successor, was swept out of power by a military coup that put an end to an otherwise decaying and unsteady forty-eight-year-old dictatorial regime. The return to democracy was lavishly praised at the NATO's foreign ministers' meeting at Ottawa the next June. At that meeting the new foreign minister, Mário Soares, pledged, with warmth and a great feeling of identification, the new regime's full acceptance of its obligations under the treaty: "For the first time in its History does . . . the Organization hail a Portugal fully respectful of the values emblazoned in the Preamble of the treaty. . . . Were it to be done now, Portugal would be ready to sign it in an attitude of full loyalty and good faith."[28]

The congratulatory feelings so earnestly expressed then were somewhat undercut the following year when Portuguese politics underwent a sharp turn to the Left. NATO was thereby given cause to put in serious doubt Portugal's trustworthiness. It seemed to be the intention of the Portuguese government to stay in NATO to maintain the balance of forces in Europe. On the other hand, the Communist elements, sometimes disguised, expressed sympathies and affinities with the Warsaw Pact countries, accompanied by loud, deprecatory rhetoric against capitalism and the "imperialist" West. In Brussels, the big question was whether or not NATO could rely on such a partner in case of actual confrontation with the powers of the Warsaw Pact, or even in day-to-day relationships. It was later said that Leonid Brezhnev himself had advised his new friends to remain in the alliance; the country was in the NATO area, and it could not be otherwise. Since the leftist movement came to its inglorious end on 25 November 1975, relations with NATO have been normal, although of late marred by some passiveness in responding to Portuguese requests for military aid.

PORTUGAL IN NATO

Some observers seem not to realize even yet that 1974 marked the end of one era in Portugal and the beginning of another. This is so not necessarily because dictatorship was wiped out and democracy restored, which in itself

[28] Text from classified records in the Portuguese Ministry of Foreign Affairs, Lisbon.

was a landmark, but because of the tremendous impact of decolonization. The era of expansion had to give way to the era of retrenchment, and with it came the need for reevaluation of a number of positions and national purposes. The historic reason for Portugal to rely on the "services and utilities" of the British alliance in modern times—to keep open the routes of the colonies under the protection of the world maritime power—has faded away. For many Portuguese the alliance may have lost its usefulness by the simple fact that Portugal no longer has colonies and needs no major sea lane protection.

From a strategic point of view, Portugal is defined by a triangle reaching from a point on the continent of Europe to the two areas where the archipelagos of the Azores and Madeira are located. Some strategists call the whole an "archipelago," or a "quasi-archipelago," a conception somewhat authenticated by NATO when it excluded the mainland from SACEUR and included the whole in SACLANT, under which a regional command, IBERLANT, is located on the outskirts of Lisbon. The maintenance of Spain's presence in NATO probably would reinforce the different strategic roles envisaged for each of the two countries: Spain would be the Iberian component of European continental defense and Portugal would provide the Iberian component of Atlantic defense. A mere glimpse at a map reveals the high strategic value of the three components of Portugal, which together are at the crossroads of the main Atlantic routes on which Europe is highly dependent.

> [T]his strategic location in a maritime area through which a substantial part of the reinforcement of Europe would pass in an emergency makes the so called 'Portuguese triangle' a possible fulcrum for:—support operations for rapid reinforcement of Europe;—control of the approaches to Gibraltar;—support of the southern boundary of Europe and to operations of containment of any eventual threat coming from the south.

Moreover, on their side, the Azores and Madeira

> constitute vital support for bases for monitoring of Soviet surface and sub-surface forces in peacetime, and in wartime they would be essential as bases for staging reinforcements in transit to Europe and for the monitoring of combat operations. Quite recent events have again shown the importance of the Azores in a scenario of a local out-of-area conflict in an area with such an importance as the Middle East.[29]

As to the specific case of the Mediterranean, it should be recalled that, among the southern members of the alliance, Portugal is the only one not actually bordering on the sea. Nevertheless, the Portuguese geographic position could give a unique and substantial contribution to any allied effort in a situation of tension or conflict in the area. All the Atlantic routes converging

---

[29]A. de Sousa Leitao, "The Portuguese Navy and the Portuguese Strategic Triangle," *NATO's Fifteen Nations,* Special Issue 27 (February 1982): 72.

to the Strait of Gibraltar are entirely under the combined surveillance capabilities of the three Portuguese components: the southern route by Madeira, the eastern route by the Azores, and the northern route by the mainland. In an emergency one of these can substitute for the others. This example of potential participation in strategies of attack, interdiction, and dissuasion is ample evidence of strategic relevance. It might be added that the country's neighbor to the south, Morocco, has an Atlantic coastline and is located in an area whose surveillance can be assured by the Portuguese mainland and by Madeira.

In speaking of the strategic position of Portugal, it is clear that there is a need to change, or at least to revise, some worn-out concepts, one of which is "geostrategic periphery," a term generally associated, for instance, with the ideas of reserve, rear guard, bridgehead, and immobile aircraft carrier. In that sense Portugal is a paradigmatic case. By the same token, the Azores and Madeira also possess those qualifications. All this is true in terms of geography but may be misleading in terms of strategy. The hard fact is that no peripheral positions exist any longer in the Euro-Atlantic NATO areas.

Strategists less frequently speak today of the United States as the sole world maritime power and more often of the Soviet Union as one of the two maritime powers now established in the world. The Portuguese writer, Virgilio de Carvalho, felt compelled to warn that "it is hardly conceivable to consider the [Portuguese] territory only, or even mainly, as a rearguard, or a disembarkation quay in Europe . . . since it might well happen that, in the strategy of the Euro-Asian super-power—now with enough capability to make use of the sea—[those] positions will be after all in the front line." Therefore, the enemy may arrive first to annihilate instead of coming later to interdict allied utilization of positions.[30] Earlier in his thought-provoking article, de Carvalho wrote of a Soviet "strategy of systematic demolition and penetration of areas of traditional influence and support for the Maritime [U.S.] Power in the periphery of the Euro-Asian Island." He also mentioned another maneuver that the Euro-Asian (Soviet) Power may attempt, one aimed at the separation of the Euro-Western allies from the Americans through the use of forces acting by proxy; for example, the Algeria-Libya-Mauritania-Polisario situation as it existed in 1981.[31]

These are, however, military considerations which, by definition, comprehend only particular aspects of a general strategy. For decades the Soviet Union has been pursuing, with determination and a timely sense of opportunism, a gigantic operation that might be considered a modern version of what was once called containment. The Soviets seem to be developing an outer containment of their potential NATO enemy. Their goal is not just to build a ring around the area, or areas, where the potential NATO enemy is

---

[30]See also Virgilio de Carvalho, "A Terceira Guerra Mundial e as ameacas a Portugal," *Nacão e Defesa* 17 (January–March 1981): 166.

[31]Ibid., pp. 154–55.

concentrated but to encircle it in order to cut off access to raw materials and oil resources. That policy is being implemented continuously when feasible, or intermittently when necessary, piercing here and there or responding promptly to opportunities extending down through the Middle East to the whole of Africa and continuing on to the Atlantic islands (independent countries), Latin America, and elsewhere whenever possible. Portugal thus stands as an important barrier to Soviet movement into the south Atlantic as well as one of the guarantors of NATO's control of the Western entry into the Mediterranean.

Spain's Entry Into NATO*

VICTOR ALBA

Four circumstances made Spain's entry into NATO in spring 1982 a special case:

1) It is the first country to join NATO since 1955. It therefore is in the position of dealing with a working institution without having the advantage, as did the founding members, of being able to influence its structure.

2) Its joining did not take place either during a Cold War period, as was the case when the alliance was founded, or during an age of détente, as was the case during the following years, but rather during a time of hesitation, disorientation, and absence of specific international policies and clearly defined objectives.

3) Spain joined NATO after many years of dictatorship, during which it was somewhat shunned by the organization. Therefore, the Spanish expect more from the alliance than it probably can offer, such as protection against right-wing coups, in spite of the examples to the contrary in the instances of Greece and Turkey.

4) Outside of the context of NATO, the United States maintained special relations with the Spanish dictatorship. The anti-Americanism in democratic Spain that resulted from those relations has extended to the alliance.

FRANCO AND NATO

In view of the vigor with which Franco affirmed his anticommunism, Spain should have been among the founding members of NATO. Nevertheless, it has only been since the generalissimo's death that the possibility of Spanish membership in the Atlantic Alliance has existed. Indeed, the very nature of the Franco regime and the international reaction that it aroused were obstacles which the member governments would have had difficulty in surmounting, even if they had been tempted to bring Spain into NATO in 1949.

From 1953 on the military policy of the Franco government was built upon a treaty of mutual aid and cooperation with the United States. Spanish national defense, in fact, was dependent upon the American bases built in Spain under the terms of that treaty. But in spite of the treaty, or at certain

*Translated by Mark R. Rubin.

times because of it, it was in the interest of Washington for Spain to belong to NATO. On the one hand, the western flank of the Mediterranean was becoming weaker as the European member nations lost their colonies, and on the other hand, it was becoming more and more costly to maintain the bases.

When the alliance was founded, there was no support for a Spanish contribution outside of Portugal. The Washington-Madrid treaty in 1953 was the Franco regime's first step toward respectability. After that John Foster Dulles began making habitual stopovers in Madrid when returning from NATO meetings in order to keep Franco apprised of developments. In 1958 a comprehensive plan for the modernization of the Spanish army was adopted in order to seduce the member states of the alliance with the promise that Spain, which until then had been militarily inconsequential, would be able to make a contribution.[1] When the Spanish-American treaty was signed, it was understood that it was not a prelude to NATO membership but rather an alternative to it.[2] Nevertheless, from 1955 on Dulles affirmed that the United States would view membership favorably.[3]

The Franco regime took care not to insist publicly on NATO membership. At first the Spanish press, entirely controlled by the government, tended to reduce the importance of the alliance. The U.S. role in Spain's economy, as well as in its defense system, compensated in part for exclusion from the organization. In the words of Senator Allen Ellender in 1957, "we were led to believe . . . that the base construction in Spain would prime the economy of Spain. However, now we are requesting aid grants for Spain."[4] Then the official position changed; Spain would not take the initiative of asking for membership, but, if an offer were to be made, it would be considered.

The invitation did not materialize, in spite of the efforts of Lisbon and Washington, the favorable disposition of the Conservative British government (which had to abandon it in the face of protests from Labour opposition), and later on of General de Gaulle. Finally, the German conservatives took up the cause and in 1960 examined the possibility of establishing a special Hispano-German military liaison that would allow the stationing of Spanish forces in the Federal Republic of Germany. That project was finally dropped when the Social Democratic-Liberal coalition came to power at the end of the decade. Socialists of all countries were supported in their refusal by the vetoes of Belgium, Denmark, and Norway, either for religious reasons (the discriminatory treatment imposed upon Protestants in Spain), or for reasons having to do with the Second World War; for example, Spain had refused to extradite Léon Dégrelle, the Belgian Fascist leader wanted by the Brussels government.

---

[1]Arthur P. Whitaker, *Spain and Defense of the West* (New York: Harper, 1961), pp. 75–77.

[2]James W. Cortada, *Two Nations over Time: Spain and the United States, 1776–1977* (Westport, CT: Greenwood Press, 1978), pp. 230–33.

[3]Ibid., p. 232.

[4]Whitaker, *Spain and Defense of the West,* p. 78.

The final attempt was particularly clumsy. Fearing that after the 1974 revolution Portugal would withdraw from NATO or that there would be Communist ministers in the Lisbon government, Washington expressed an increasing interest in a Spanish connection with NATO in May 1975.[5] Meanwhile, however, the Franco government had adopted a harder domestic line, thus provoking increased criticism in Europe; all of the other alliance members used their vetoes. Once more Washington's image deteriorated: in anti-Franco public opinion because of its toleration of the Fascist leader, in pro-Franco public opinion because of its inability to impose its views on NATO.[6] Here it is necessary to realize that opposition to dictatorship should not be characterized by realism, and that it is usually as ill-informed as the general population to which it belongs. The American information services in Spain did not broadcast anything that would embarrass or irritate the dictator. They thus reinforced the lack of realism and contributed to the fact that the Franco regime's permanence was commonly attributed to American aid and not to Franco's cleverness or to his opposition's weakness. It is important to an understanding of contemporary events to recall that many positions in the government and in the political parties, as well as seats in the parliament, are occupied today by those who were in opposition to Franco and who inevitably share that vision of American foreign policy toward Spain.

In 1975, while Franco was still alive, a public opinion poll revealed that 56 percent of Spaniards were against the presence of American bases on Spanish territory, and that 33 percent claimed to be in favor of a policy of neutrality. Even under Franco there were challenges to these bases in the Cortes, although the body was thoroughly domesticated and its members were named by the state.[7]

THE U.S. IMAGE IN SPAIN

It is almost impossible to understand the circumstances of Spain's entry into NATO without placing them in the perspective of Hispano-American relations. The majority of Spaniards concerned by this matter identify NATO with the United States.[8]

Anti-American sentiment in Spain, especially in the upper classes, dates from the Spanish-American War of 1898 and from the role that the United States played in it. That same sentiment spread to other social classes because

---

[5]See Pierre Hassner, *New York Times*, 30 May 1975. See also report of meeting of NATO defense ministers, 22–23 May 1975, in *Facts on File* (31 May 1975): 373.

[6]James W. Cortada, ed., *Spain in the Twentieth-Century World* (Westport, CT: Greenwood Press, 1980), p. 104.

[7]Eduardo Chamorro and Ignacio Fontes, *Las bases norte-americanas en España* (Barcelona: Editorial Euros, 1976), pp. 239–40.

[8]See Cortada, *Two Nations*, pp. 230–33.

of President Franklin Roosevelt's neutrality during the Spanish civil war (embargo of armaments to both sides but not of petroleum to the Francoists). It intensified still more as a result of the lukewarm attitude adopted by Washington toward the plan to overthrow Franco after the Second World War; the American vote to admit Spain into the United Nations (while forgetting a similar Soviet vote); the Hispano-American treaty; and visits to Franco by Dwight Eisenhower (December 1954), Richard Nixon (October 1970), Spiro Agnew (July 1971), and Gerald Ford (May 1975), as well as William Rogers and Henry Kissinger's refusal to meet with opposition delegates during their visits to Madrid.[9] If we add to that the collaboration of the CIA (ended by President John Kennedy) with Franco's secret police, which allowed the latter to know about the opposition's overtures to the American embassy and even to arrest those who were undertaking them, we can understand that opposition by a segment of the Spanish population to NATO was predetermined. Both the Washington and Madrid governments proved incapable of dissociating them in the mind of the Spanish public.

In short, many Spaniards of all political tendencies believed, and continue to believe, that the special relationship between the Franco regime and Washington had saved the former when it was in difficulty. Common sense, though, indicates that such a view is erroneous. Nonetheless, the real influence of the United States in Spain diminished rapidly because it depended less upon pressures and diplomacy than on popular opinion and support.[10]

The accidental fall of a nuclear bomb from an airplane off of the Spanish Mediterranean coast on 17 January 1966 increased the irritation of the Spanish public opinion toward the Hispano-American treaty. The American ambassador and a Spanish cabinet minister even bathed in the supposedly contaminated waters but to no avail.[11] In 1975 a collection was organized at the U.S. military base in Rota to establish a fund that would honor Francoist repression forces with a cash prize. Such a gesture could surely not be expected to increase popular support for the existence of those bases, nor did Washington make friends in Spain in October 1975. While all of Europe was condemning Franco following a series of political executions, while ambassadors were being recalled, and while Spain remained virtually isolated, Kissinger took

---

[9]Cortada, *Spain*, p. 249.

[10]This fact is evidenced in Carlton J. H. Hayes, *The United States and Spain: An Interpretation* (New York: Sheed and Ward, 1951). The realism of American diplomats, obsessed by this entelechy that they call "stability," is lacking in realism. In his book, Hayes affirms that "collaboration with Spain requires, of course, an overcoming of democratic scruples about General Franco's government, which is undoubtedly a kind of dictatorship." And he prophesies: "Besides, what would be the alternative to the present regime in Spain? . . . The alternative would be a renewal of domestic strife between 'Leftists' and 'Rightists' . . . and a consequent recurrence of violence and civil war, ending in another kind of dictatorship worse than the present" (pp. 181–83). Suffice it to say that after Franco's death the transition from dictatorship to democracy was accomplished with neither violence nor civil war.

[11]Flora Lewis, *One of Our H-Bombs Is Missing* (New York: McGraw-Hill, 1967), pp. 176–79.

advantage of the situation to obtain concessions from Franco on the renewal of the bilateral treaty. He did not condemn the regime's behavior or appeal for "clemency" on behalf of those executed, as some European leaders had.[12]

To this list of grievances should be added those arising from the attempted military coup of 23 February 1981, when democracy was almost overthrown in Spain. These are minor irritations, but, in the context of the past record, they take on an importance that admits of no ambiguity to a Spaniard, even though they seem exaggerated to Americans. What are these grievances? They are built on rumors. The Spanish military attaché in Ankara, Colonel Federico Quintero, wrote a favorable report on the Turkish military coup. That report was circulated among the military men who subsequently revolted in Spain. The most damaging part of this whole affair was the allegation that it was written at the suggestion of the CIA zone chief, John H. Kenny. When he learned of the attempted coup, Secretary of State Alexander Haig said only that it was an "internal Spanish matter." For his part, the American ambassador to Madrid, Terence Todman, confided to a reporter several days later: "Don't forget that the United States is a friend of Spain, during the Franco era and during that of King Juan Carlos." Up until the morning of 24 February, when the surrender of the rebels was already being negotiated, the Washington-Madrid "hotline" remained quiet. The explanation for this long delay is that the White House was waiting for President Ronald Reagan to wake up, and that as soon as he was briefed, he called the king.[13] He was the last head of state to do so, hours after Helmut Schmidt, Valéry Giscard d'Estaing, Margaret Thatcher, Alessandro Pertini, and even King Hassan II of Morocco.

Given the apparent insensitivity of the United States to the security of Spanish democracy, it would thus seem logical that when Madrid decided to enter NATO the bilateral character of U.S. bases in Spain would change. No such thing occurred. Madrid persisted in negotiating the renewal of the treaty. First, because it received a number of advantages from it such as arms or credit for its military forces. Second, it was feared that, if the bases were to be made part of NATO's infrastructure, Washington would stop maintaining them, and that the cost of their upkeep and operation would fall in large part on Spain, even if NATO should assume a portion of the cost.[14]

---

[12]*New York Times*, 12, 18 March 1981.

[13]Pilar Urbano, *Con la venia . . . yo indagué el 23 F.* (Barcelona: Argos Vergara, 1982), pp. 25–26. The fact that several of the rebels involved in the 23 February 1981 coup attempt had taken courses in American military schools, and that one of the key figures in the coup, Lieutenant Sánchez Valiente, is able to live peacefully in Texas while he is wanted by the Spanish military authorities, has certainly not embellished the image of the United States in Spain (pp. 136, 140).

[14]Cortada, *Two Nations*, pp. 262–63. The new agreement, signed on 2 July 1982, recognized Spanish sovereignty over the bases and prohibited both landings at them and flights over them of nuclear materiel, without Spanish authorization; moreover, it made provisions for $415 million in military aid.

THE REASONING OF THE OPPOSITION

In addition to anti-Americanism, the opponents of membership in NATO had on their side what may be called "ancestral memory." Indeed, during its imperial period Spain was involved in European affairs, often was at their center, and participated in European wars. One can even argue that Spanish decadence originated in this "continental intervention." Be that as it may, Europe appears, in the collective Spanish memory, as the source of the country's ills. The civil war itself is viewed as a transposition to Spain of a European phenomenon, fascism, and because of the latter, of another European phenomenon, Stalinism. Spain's entry into NATO, therefore, appears to be a return to the source of its misfortunes.

The Marxist Left reinforced anti-NATO sentiments from a different perspective. In spite of its criticism of certain Soviet "excesses," such as Afghanistan and a few years ago Prague, the Spanish Communist Party supported the fundamental principles of Soviet diplomacy. The splinter groups that have broken away from the mainstream—Trotskyists, Maoists, Stalinists, and even terrorists—are opposed to NATO membership as a matter of dogma and because they do not want to be associated with the party's "right wing."

This ideological opposition has little specific weight. The same is true of other groups: pacifists, ecologists, and those who feed upon leftover literature of the old "new left" and who persist in maintaining the same state of mind as when they were in opposition to Franco. The significant opposition—the only significant one—is that of the Spanish Socialist Workers' Party (PSOE), the most important political party in Spain. The motivations of this opposition are not found in the arguments put forth by it, which are tactical and not strategic. They derive from a mixture of distrust of the United States, a desire to attract the support of the admittedly small but numerous groups whose combined votes could mean the difference between winning and losing an election, and from the desire to avoid creation of obstacles to the possibility of future support of a Socialist government by the Spanish Communist Party.

Aside from political opportunism, there subsists something profound deep within the history of the Spanish Socialist movement. Spanish Socialists, who were opposed to the Third International in 1919–21, had put their confidence in President Woodrow Wilson's Fourteen Points and had made them the cornerstone of their foreign policy. They believed that they had been betrayed by Washington when the Senate refused to approve American membership in the League of Nations. Since that time the PSOE has always distrusted the United States. It might be added that many of the present Spanish Socialists began their political careers in the opposition as Communists, and, although they quickly left the Spanish Communist Party, they still retain a certain number of the most facile prejudices and reactions: anti-Americanism and one-way anti-imperialism. Moreover, it should be noted that, while maintaining their verbal opposition in internal politics, the Socialists collaborated with the centrist government, spurred, no doubt, by the fear of a military coup, and that the NATO question provided them with the chance to differentiate themselves from the government on an issue that did not directly affect internal affairs.

ARGUMENTS AGAINST NATO MEMBERSHIP

Many NATO opponents have not come out against membership as such, nor have they come out in favor of Soviet diplomacy. Their argumentation is mainly centered on two points, one of which is a constitutional question. They claim that, in view of the importance of membership for the country's future and the effect that it would have on all Spaniards, it should be put to a referendum. A petition campaign to force such a referendum was instigated by both the Communists and several pacifist groups. It was quickly abandoned and followed by a warning from the Socialists that they would organize one if they came to power. According to the second argument, Spain's membership would disrupt the East-West balance of power and increase the chances of new conflicts, which would be harmful for Spain and for the rest of the world. Because it would reinforce the position of NATO, membership would speed up the arms race.

Two other arguments complete the dossier: Would it not be better to wait to join until the defense of Europe is completely in the hands of the Europeans, free of all dependence on the United States? And can Spain afford the considerable expense connected with alliance membership during the present economic crisis?[15] It is worth observing that most opponents of NATO membership never proposed abrogating the military treaty with the United States. To the contrary, they believed that it was sufficient for the national defense and saw no necessity for joining NATO to consolidate Spain's territorial security.[16]

These opponents have cited the cases of Greece and Turkey to demonstrate that membership in NATO is not a "vaccination" that would render a country immune to military coups. They have stressed, moreover, that in view of its functions the alliance would be of no help whatsoever if a conflict arose with Morocco over Ceuta and Melilla. Finally, they have maintained that, since Britain is more powerful than Spain, NATO would always support the United Kingdom on the issue of Gibraltar; consequently, membership would restrain Spain's ability to maneuver.[17]

---

[15]In his book, *OTAN y España: El precio de una alianza* (Madrid, Sábado Gráfico, 1981), Angel Lobo concludes that these expenses would represent 0.31% of the Spanish defense budget.

[16]Fernando Morán, *Una politica exterior para España* (Barcelona: Planeta, 1980), pp. 84–114. The author is a Socialist leader and a member of the Spanish diplomatic corps. He is now Spain's minister of foreign affairs.

[17]Felipe González, the present head of the Spanish government, has said, as secretary general of the PSOE, "we are for NATO, but against Spain's entry into NATO." *Informationes* (Madrid), 20 January 1979.

SOVIET WARNINGS

Franco's Spain maintained frequent contacts with Soviet bloc countries. As early as 1946 the British delegate to the United Nations denounced Spanish trade with Poland. Commerce between Spain and Communist countries, sometimes through the intermediary of Arab countries, had never stopped. In 1969 and 1970, at the same time that miners in the Asturias were carrying out widespread illegal strikes, Poland sold considerable quantities of coal to Spain to make up shortages caused by the strike. The Spanish government also authorized two Soviet ships to anchor in the waters of the Canaries—a strategic location—in order to study the possibility of establishing a fishing installation there. Regularly scheduled air traffic now operates between Madrid and the Communist countries, and the latter, including China, have established agencies in the Spanish capital. In 1972 a Hispano-Soviet commercial treaty was signed. Diplomatic relations with East Germany were established in 1973 and shortly thereafter with China.

The USSR was pleased by Spain's support of its proposal to convene in 1980 a European security and cooperation conference; in exchange, the Soviets seconded Madrid's demands in the United Nations for the decolonization of Gibraltar. The Soviet government did not favorably view Spain's entry into NATO, but it had no way of preventing it. Trade between the two countries was of minimal importance, and neither the Spanish Communist Party nor the Communist unions had adequate leverage. It could count, however, upon the anti-American feelings held by a portion of the Spanish population and upon the latter's fear of being pulled into a European war. According to a number of observers, the Soviet's best weapon was that of indirectly fomenting terrorism by means of secret aid funneled through Libya and other countries in order to provoke a military coup that would close the alliance's doors to Spain.[18] But when that did not suffice, Moscow had to be satisfied with threats.

As membership in NATO began to be discussed seriously in Spain, an editorial in *Izvestia* underlined that such a move "would affect the desirable balance [existing] between the two military-political organizations."[19] In September 1979, Castro asked, in his opening address of the Sixth Conference of Non-Aligned Nations, that Spain not allow itself to be drawn into the aggressive NATO bloc.[20] During a visit to Madrid several months earlier, and during talks with high-government officials, Soviet Foreign Minister Andrei Gromyko had steered talks in the same direction. *Pravda* and the TASS official news agency maintained that Spanish membership would contravene the final declaration of the Helsinki Conference, that it would impede

---

[18]*El Noticiero Universal* (Barcelona), 17 August 1981.

[19]Cited in *El País* (Madrid), 4 April 1978. Similar declarations from Andrei Gromyko cited in *Avui* (Barcelona), 20 January 1979, and from Afanasiev in *El País*, 22 April 1978.

[20]*El Noticiero Universal*, 4 September 1979.

understanding between the two countries, and that it would "preoccupy supporters of détente, of peace, and of the security of nations."[21] On 7 July 1980, according to the German daily, *Die Welt*, Soviet President Leonid Brezhnev maintained, in conversations with Schmidt in Moscow, that membership "would worsen the situation in Europe" because it would demonstrate that NATO aspired to extend its field of action.[22] The same message was received by Willy Brandt, president of the Socialist International (El Salvador), during his visit to Moscow in July 1981.[23] Also, according to a Spanish weekly, a Soviet leader proposed aid, in the fight against the ETA and its terrorism, to Marcelino Oreja, minister of foreign affairs, if Spain would stay out of NATO.[24]

On 7 September, Oreja received a Soviet memorandum. The Spanish minister, considering that it constituted unacceptable interference in Spain's internal affairs, sent it back. In addition to the arguments already presented, it said that "the inclusion of Spain in an alliance aimed against the USSR and its allies could not fail to affect relations between the Soviet Union and Spain," and that "the USSR and its allies would be forced to draw suitable conclusions and to weigh the possibility of adopting appropriate measures." Finally, it was stated in the memorandum that if Spain entered the alliance it would no longer be considered as a "country not possessing nuclear weapons." In May 1982 Spanish diplomats and journalists in Moscow reported that they were suddenly being treated with marked coolness. Spanish diplomats also were made to understand that, if their country abandoned membership in NATO, the USSR would be favorably disposed to conclude a nonaggression treaty, in which Moscow would promise never to use nuclear arms against it.[25] In July the Spanish press reported a campaign of disinformation organized by the KGB to prevent Spain from joining the alliance and to prejudice the prestige of the United States in the eyes of its European allies. Among the falsified documents was a letter from President Reagan to the king of Spain, copies of which already had been circulated during the European Security and Cooperation Conference at Madrid in November 1981.

Of all the Communist countries, only China diverged from Moscow's line. It had approved of the reestablishment of the monarchy in Spain (the king made an official visit to Peking in 1978) and supported Spain's request for membership in the European Economic Community (EEC) and in NATO, both by commentaries in its press and by official statements.[26]

The Franco regime, by blaming all domestic opposition on the Communists and by carrying on a narrow-minded anti-Communist propaganda campaign, prevented many Spaniards from discovering the true nature of the Soviet system. There are many who still believe that the USSR is a peaceful

---

[21]Cited in *El País*, 22 April 1978.
[22]Cited in *Avui*, 9 July 1980.
[23]*La Vanguardia* (Barcelona), 14 July 1981
[24]*Sábado Gráfico* (Madrid), 16 August 1981.
[25]*La Vanguardia*, 20 May 1982.
[26]Cortada, *Spain*, pp. 26–28.

power whose promises may be taken at face value, and that a war could only originate with the United States, never with the Soviets.

REASONS FOR SUPPORT OF NATO MEMBERSHIP

The moving force behind membership was the Central Democratic Union (UCD, a centrist party), which also was responsible for the transition from Francoism to democracy and which governed the country from 1977 to 1982. The Popular Alliance (AP, the rightist democratic party) and the centrist Catalan nationalists, who govern in Catalonia, were favorable to NATO as well. The consequent polarization caused by the membership issue made it appear to public opinion as a dispute between the Left and the Right. That perception will undoubtedly hang over the future of Spanish collaboration with NATO.

It was also the UCD that exerted pressure to have the NATO question taken under consideration and resolved during the legislative session that ended prematurely in August 1982. But even within the party, there were those who thought that the problem was not urgent, that it should be debated until a point of saturation was reached sufficient to allow for approval of membership without great trauma. Nonetheless, the UCD, and especially the Calvo Sotelo government, which was constituted immediately after the attempted coup of 23 February 1981, successfully insisted that the membership project be presented immediately to the parliament.

It is possible, in fact probable, that the urgency was due to pressure from the United States but less so, perhaps, than claimed by opponents. Aside from this pressure, two assumptions no doubt moved the UCD to resolve this problem quickly. First, with entry into NATO the chances of a military coup would lessen (this is one of the arguments made in *private* conversations; the Spanish military could possibly be offended). Second, public opinion polls showed that the UCD had little chance of winning the next election. Moreover, that union was splitting up. Social Democrats, Christian Democrats, and Liberals (three groups that at the outset had founded the UCD, along with the democratized Francoists) had broken away to form their own parties. That process was already under way when Prime Minister Calvo Sotelo presented the membership project to parliament, but it had not yet had the effect of reducing UCD parliamentary votes. For the moment, the votes of the UCD, the AP, and the Catalan nationalists were assured and meant parliament's approval of NATO membership. Nothing would guarantee that result in a new legislature. Inasmuch as the UCD believed that membership in NATO and in the EEC (the latter enjoying broad support) was an essential basis for the democratization of Spain and for the stabilization of its political system, it wanted to lay these foundations for the future.

ARGUMENTS IN FAVOR OF MEMBERSHIP

While the adversaries of NATO membership were plastering walls with posters and multiplying demonstrations, although the principal opposition—the PSOE—was quite inactive on this front, its proponents had not organized any public campaigns. They were content to defend their position in the parliament and in their press releases. The proponents' main arguments were that membership, along with democracy, is the logical consequence of Spain's Western and European option, and that the geopolitical situation of the country makes neutrality impossible. According to Sánchez Gijón, membership in NATO rationalizes Spain's new role in the West: maintaining free access to raw materials; securing petroleum at a price determined by market conditions and not by political or military considerations; and preventing the most important continental land power, the USSR, from becoming the most important naval power as well.[27]

As for its geostrategic importance, Spain constitutes a rearguard position of support for the Rhine, the Po, and the Loire. It is also a forward American port in Europe and the European antechamber of the South Atlantic routes. Finally, it occupies a very special position in the Mediterranean. The existence of the Hispano-American treaty, until the present, had limited Spanish strategic notions to the Madrid-Washington axis, inhibiting the perspectives of a European strategic axis vital to the defense of Spain. Previously, in the event of conflict, the country would only have been a platform for American forces. But once in NATO, Spain and Portugal could coordinate their zones of common interest and even succeed in establishing, within the alliance, an Iberian Command (similar to the Allied Command Channel) in which, for strategic reasons, Gibraltar in some way would be integrated.

Opponents of membership evoke the examples of Sweden and Switzerland to buttress their case for neutrality. But in view of its geographic position, Spain cannot be compared to those countries, say partisans. As a strategic bastion, it should perhaps be compared with Iran and Turkey. Finally, supporters maintain that it would be absurd to join the EEC without joining NATO, since the existence of the latter makes possible the working of the former.

THE MILITARY POSITION

To understand the position of the armed forces, it should be recalled that in Spain, as in Latin America, militarism—military intervention in politics: coups, dictatorships, etc.—has liberal origins. Immediately following

[27]Antonio Sánchez Gijón, "España y la Opción atlántica," *Historia* 16 (October 1980): 58–64. The author is the secretary of the Instituto de Cuestiones Internacionales in Madrid.

the war against Napoleon, when absolute monarchy was reestablished in Spain, no means of legal expression remained for the common people who had borne the weight of the war. Many guerrilla leaders of lower- or middle-class origin had distinguished themselves in the struggle. Once integrated into the regular army, they used their commands to attempt to overthrow the absolutist regime. Then the civil wars of the nineteenth century gave an inordinate importance to the army. In 1868 a liberal military coup ended the reign of Queen Isabella II. Another military junta, also liberal, reestablished the monarchy in 1876.

In the next generation the army conducted an unsuccessful colonial war in Cuba and another in Morocco. Those humiliations helped to create a suitable atmosphere for the 1923 military coup of General Miguel Primo de Rivera, in complicity with King Alfonso XIII. But in 1930 this same monarch dismissed the dictator and attempted to return to a constitutional system. He was overwhelmed by popular distrust, as manifested in the 1931 municipal elections in favor of the republic. Once it had been proclaimed, a modest military reform was carried out, consisting of voluntary early retirement made necessary by the fact that there was one officer for each unit of four soldiers. Nevertheless, the greater part of the army remained monarchist and opposed to regional autonomy, which it considered to be a factor of the disintegration of "national unity." In August 1932 an attempted coup failed, but in July 1936 another was eventually successful after thirty-three months of civil war.[28]

General Franco, who had made certain of the army's support before beginning his counterrevolution, kept the military out of politics during the forty years of his rule. It began as a Fascist regime and then, after World War II, became a "traditional" dictatorship. There were, assuredly, a few conspiracies seeking to restore the monarchy, but the armed forces remained loyal to the caudillo. He manipulated military and personal rivalries in order to keep the army out of politics and was aided in that endeavor by his tolerant attitude toward administrative corruption and constant favoritism toward the military. Nonetheless, the army continued to be poorly paid and was not particularly well armed.

When Juan Carlos II took the throne after Franco's death, he was able to count on the support of the armed forces; he had even studied in the respective academies of all three military branches. However, the young king gave rise to surprise and discontent among older military men by endorsing a constitutional democratic regime instead of perpetuating Francoism. That discontent, fed by civilian partisans of Francoism who were angry about losing official support for their long-standing corrupt practices, was manifested by

---

[28]On the question of Spanish militarism, see Stanley G. Payne, *Politics and the Military in Modern Spain* (Stanford: Stanford University Press, 1967); Miguel Alonso Baquer, *El ejército en la sociedad española* (Madrid: Edicións del Movimiento, 1971); Manuel Cabeza Calahorra, *La ideología militar hoy* (Madrid: Editora Nacional, 1972); "Bellius," "Les militaires et le développement démocratique," *Le Monde diplomatique* (Paris), September 1979; and Colectivo Democracia, *Los ejércitos más allá del golpe* (Barcelona: Planeta, 1981).

a number of plots—discovered in time—in 1978 and 1979, and by the attempted 1981 coup. Because of his authority, the king was able to keep the army from supporting the rebels.

The debates over Spain's entry into NATO took place, therefore, in an atmosphere of unrest over the possibility of a new coup, in spite of the placement of younger officers in high commands. It does not, however, necessarily follow that young officers have a more modern mentality. The instruction that they receive in their schools is dispensed by professors trained under Francoism. Furthermore, a persistent argument of NATO partisans was that membership would reduce the risk of a coup. And, it is precisely for that reason that, going beyond their self-styled anticommunism, older military men and their civilian instigators only accepted membership reluctantly. But the armed forces, as a whole, have taken a position in favor of Spanish participation in NATO.

While the military has brought pressure to bear, sometimes in a powderful and even threatening manner, on such issues as divorce, regional autonomy, and legalization of the Communist Party, it never did anything comparable with NATO. The military was apparently willing to accept whatever the civilian government decided. However, that may not be the case if in the future Spain were to decide to leave the alliance. The pressures against withdrawal probably would be very strong.

NATO's POSITION

NATO's position on Spanish membership, as reflected in the international press, has been one of essential indifference. At first there was perhaps some fear that Spain would cause more difficulties than it would resolve, or that it would place the alliance in embarrassing positions on such issues as Gibraltar, Ceuta, and Melilla. In any event, NATO continued to reject the Spanish diplomatic establishment's attempt to link membership with the EEC, Gibraltar, and the security of Ceuta and Melilla. But faced with a Madrid government anxious to enter the organization, Brussels offered a vocal welcome. One month after the death of Franco, the secretary general of NATO, Joseph Luns, declared during a ministerial meeting in December 1975 that it was too early "to draw conclusions" about Spain.[29] During that same month, after the NATO ministers' meeting in Oslo, Luns stated that it was necessary to await "the putting into practice of the plans for democratization in Spain."[30]

In August 1976, General Alexander Haig, then SACEUR of NATO, visited King Juan Carlos who was vacationing in Majorca, thus inspiring speculation about the demise of the Hispano-American treaty in the event of

[29]*La Vanguardia*, 11 December 1975.
[30]Ibid., 19 May 1976.

Spain's joining NATO. After an interview with President Ford on 15 September 1976 in Washington, Luns explained that all the members of NATO approved of the treaty between the two countries and were pleased about the process of democratization in Spain. Luns said that Warsaw Pact countries' claims that the military organizations of the two blocs should not admit any new members were absurd.[31]

After eighteen months various institutes, more or less closely linked to NATO, realized that the Francoist system would not survive after Franco and began to prepare the way for Spanish NATO membership. In April 1978 Spanish politicians, both partisans and adversaries of the alliance, participated in an unofficial meeting in London to discuss the subject of "Spain, NATO, and Defense of the West." Shortly thereafter, Haig declared that NATO would accept Spain as soon as it had decided in favor of membership. From that time on, unofficial contacts were established; Spanish members of parliament favorable to the alliance went to Brussels and attended, for example, the North Atlantic Assembly on 25 May 1978. Haig's replacement by General Bernard Rogers also seemed to augur well for membership, since the new NATO commander had sat on many commissions dealing with the Hispano-American treaty. He was considered to be an expert on strategic questions concerning the Iberian peninsula.

Spanish members of parliament and senators attended the parliamentary assembly of NATO in June 1980, as observers. During the same month, at the NATO foreign ministers' meeting in Ankara, a military report was circulated in which it was stated that Spanish membership would increase "NATO's dissuasive value."[32] When the Calvo Sotelo government expressed the desire to see Spain join NATO, the press of member countries began to take increased interest. Thus, in June 1982, after an interview with his Spanish counterpart, the foreign minister of Norway (a country that had always vetoed Spanish membership during Franco's lifetime) announced support for it.[33] In Great Britain more reserve was evident, undoubtedly because of Gibraltar. The English and American press suggested in October 1981 that, if Spain joined the alliance, Spanish troops might be based in West Germany, a fact which the Spanish government had denied and which gave some ammunition to the adversaries of membership.[34]

In December 1981, after a meeting of the NATO Foreign Ministers' Council, Spanish diplomats and NATO staff members began work on the entry document. Formal admission to the alliance took place in June 1982. In spite of that event, the foreign minister of the French Socialist government declared, as late as July, that Spanish membership in the alliance would "pose

---

[31]Ibid., 16 September 1976.

[32]Ibid., 25 June 1980.

[33]Ibid., 24 June 1981.

[34]Cited in *Avui*, 24 October 1981.

significant problems for France and Portugal" and promised to support Lisbon's request that Spain be incorporated into the European Command, and that Portugal remain attached to the Atlantic Command.[35]

PARLIAMENTARY DEBATE

In his inaugural speech in February 1981, Calvo Sotelo announced his intention to propose that Spain enter NATO. In August the government asked the Council of State (the supreme judicial body of the administration) to render its opinion on the constitutional procedure to be followed. After several days of deliberation, the council issued the opinion that a simple majority vote of the parliament would be sufficient. Opponents claimed that the Council of State was politicized and that membership required a referendum. While the result of the council's decision was to reduce the problem to a question of procedure and to avoid an in-depth study of the real issue, serious wider debates would have had the advantage of preparing public opinion.

The government immediately requested the parliament's authorization to present a formal request for membership in the alliance. In September the campaign to gain public support was begun, the first on an international political subject to take place in Spain since World War I. But the expected "national debate" was limited to a parliamentary discussion, in which the fundamental question was almost always avoided. That question could be formulated thus: Does Spain believe the existence of an expansionist USSR to be a danger to its security and, if so, should it entrust its defense to others—NATO and the United States—or should it participate in a common defense of Western Europe and in the decisions made on that subject? During the debate the PSOE announced that it would appeal the procedure on the grounds of its unconstitutionality; it never did. However, the affirmative parliamentary vote did have its disadvantage. The Socialists announced that, in the event of an electoral victory, they would hold a referendum to confirm or invalidate the parliamentary decision.

According to public opinion polls taken in 1981, 69% of the Spanish were in favor of such a referendum, 3% considered membership to be a matter of high priority, 18% were for NATO, and 52% against. In the event of a referendum, there would be 44% "no" votes and 13% "yes" votes, with a large percentage of abstentions. The highest percentage of negative votes would come from Madrid, Valentia, the Basque Provinces, and Catalonia, the most highly industrialized regions of the country. Leading the "no" voters would be young people under age thirty, the upper-middle class, and men. Seventy-nine percent of those who had voted Socialist in the 1979 election

---

[35]*La Vanguardia*, 7 July 1982.

were opposed, and 32% of those who had voted for the Center were in favor. It is therefore undeniable that, given the precarious extent to which the electorate was informed, the majority was opposed to membership or, at best, very reticent about it. What was the cause of those attitudes? Five percent believed that membership would be beneficial to the national economy, 33% that it would aid in the national defense, 22% that it would be expensive, 21% that it would increase the dangers of another world war, and 21% that it would disturb the balance of power between the two blocs.[36] An opinion poll carried out in the military gave the following results: 87% of senior officers were in favor, 13% were opposed; among captains and lieutenants 47.6% were against, while 42.8% agreed and 9.5% had no opinion; and among noncommissioned officers 57% objected.[57]

After two weeks of prolonged committee discussions, the debate began in plenary sessions in the parliament and lasted for three days—from 27 to 29 October 1981. The government rejected the referendum idea as well as the inclusion of the defense of Ceuta and Melilla in the membership document, as the Socialists had requested. During the debate the Socialist representative to the Foreign Affairs Commission resigned, believing that the position of his party was ambiguous and that its opposition was "purely formal." The only concession made to the opposition—and this was to a certain extent also the wish of NATO partisans—was the inclusion in the government authorization to pursue membership, with the stipulation that it "would not accept any commitment implying the storage or deployment of NATO nuclear weapons on our territory and that, should the occasion arise, any further decisions on that subject would require the authorization of the parliament."[38]

Finally, on 29 October the vote was taken, resulting in 186 for, 146 against, 14 absent, but no abstentions. The government thus was authorized to take the necessary steps to enter the alliance.

MEMBERSHIP

On 30 May 1982 the chargé d'affaires of the Spanish embassy in Washington submitted the official membership instrument of his country into NATO, making it the sixteenth member. The admission ceremony took place in Brussels on 5 June in the presence of the council of the permanent ambassadors of the alliance. The then Spanish minister of foreign affairs, José Pedro Perez-Llorca, requested new negotiations to reconsider the Malvinas-Falkland Islands issue, a petition based on the historic bonds between Spain and Latin America

---

[36] *El País* (Madrid), 26 October 1981.

[37] Poll conducted by the military review *Defensa* in *El Noticiero Universal* (Barcelona), 27 October 1981.

[38] *La Vanguardia*, 1 June 1982.

(two weeks earlier this same NATO council had condemned the "Argentine aggression"). Moreover, the minister indicated that Spain desired the creation of a fourth NATO military command, which would comprise the Iberian peninsula, with Gibraltar, Ceuta, Melilla, and the Canary and Balearic islands. Five days later Luns confirmed that Spain had joined the alliance without conditions, and that all questions having to do with its membership would be the object of negotiations. He also stressed that Ceuta and Melilla were not in NATO's sphere and that in accepting Spain "we are aware of all the special relations which that country has with Latin America and the Arab world."[39]

The October 1982 parliamentary election gave victory to the PSOE. For the first time in the history of Spain the Socialists had an almost absolute majority in the country and an absolute one in parliament. They had constituted the essential opposition to NATO membership. During the electoral campaign the PSOE repeated that without much delay it would hold a referendum on that issue. When questioned on the subject, Socialist candidates limited themselves to the same answer: "freeze" the negotiations on military integration without withdrawing from the alliance, following the example of de Gaulle. It is thus safe to predict that Spain will remain a member of NATO, but that its armed forces will not be integrated into those of NATO unless strong military pressures are brought to bear. In point of fact, that would be maintaining the status quo. Naturally that position may be modified in one direction or the other by national or international events. In all cases, it remains evident that the entire matter of NATO is not one of the Socialist government's high priorities.

What is certain is that neither the politicians nor the military seem to have realized that Spain's entry into NATO would add yet another factor to the almost permanent tension between the United States and Western Europe. It bears repeating that until now Spain's dependence on Washington for its defense was even greater than NATO's. It is possible that one day Spain will reinforce the camp of those Europeans who argue that the defense of Europe should be the business of Europe, that Europe cannot depend indefinitely on the United States, and that the presence of two blocs within the Western alliance makes that defense uncertain, expensive, and subject to conditions imposed from outside Europe. New initiatives are needed.

It is maintained by some that during periods of crisis it is not possible to solve problems but simply to ward off the most pressing ones. Nevertheless, it is precisely during such times that the foundations for future solutions have often been laid. Roosevelt did just that with the Great Depression, Truman with NATO. This time, the initiative may come from Brussels. It will be interesting to see how Madrid, torn between Washington and Brussels, will survive.

---

[39]*Avui*, 1, 21 June 1982

The Future of Iberia and NATO

JOAQUIN OLTRA

A historian seeking to outline the course of the future on the strength of an understanding of the past is always uncomfortable. Two problems pertinent to this particular project are specially discomfiting: first, whether or not the conception of "Iberia" has specific meaning in twentieth-century history; and second, in light of Spain's strong tradition of isolationism, whether or not that nation's membership in NATO can endure for long. Each question deserves separate consideration.

THE LIMITS OF IBERIA

"Iberia" is a misleading concept. It is certainly tempting, from a purely strategic point of view, to think of the whole of the peninsula as a unified entity. The peninsula, although politically divided into three political entities, appears as a solid geographical bastion standing between the Atlantic and the Mediterranean and separated from the rest of Europe by a natural fortress of mountains. Anyone who looks at the map of Europe with military eyes cannot help wishing that the peninsula were as unified and centralized as the only other country that also sits between the Atlantic and the Mediterranean, namely, France.

Neither Spaniards nor Portuguese ever talk about their country as Iberia. The name is reserved to the whole of the peninsula and more often than not it has a purely geographical reference. Politically and culturally the term has no practical meaning, except in a very general sense as the product of a certain common history. In the Middle Ages the peninsula was a mosaic of kingdoms that by the year 1500 was finally reduced to the two countries that coexist today.

For a long time these kingdoms were five in number and very different from each other in geopolitical terms. While two kingdoms—Aragon and Granada—had a clear Mediterranean vocation, Portugal had, of necessity, an Atlantic orientation. Navarre was not a maritime power and never had the opportunity of becoming one since, in spite of its early and promising southern expansion, was soon landlocked by Castile and Aragon. The fifth, and in the long run the most influential of these five kingdoms, Castile, was ambivalent. Although it opened itself timidly toward the Mediterranean through present-day Murcia, it eventually expanded on the other side of the Atlantic. There might have been a North African as well as peninsular Castile if it had followed its perhaps more natural geographical expansion and had conquered most of the Maghreb instead of going toward the New World. As it is today, only two cities—Ceuta and Melilla—remain as a sample of what might have been.

Castile did manage, through force and in a short period of time, to subjugate two of the other five kingdoms—Granada and Navarre—incorporating their territories. Also, through a royal union the federated kingdoms of the crown of Aragon became united to Castile, even though they kept their own institutions for two more centuries, until 1715.

Portugal was then the only one of the five peninsular kingdoms that survived the ambitions of Castile. When in 1581 the Castilian Philip II proclaimed himself as king of Portugal, it was too late; the Portuguese already had a strong national identity that helped them to recover their independence in 1640. A similar attempt, also in 1640, by one of the four countries of the old crown of Aragon—Catalonia—ended in failure. The result, then, was the two independent countries which today occupy the peninsula. Yet there is still a temptation to think that there is one too many. In fact, all indications are that at least before 1500 both countries considered unification desirable and possible, as suggested by the marriage policies of the two reigning monarchies. Philip II invaded Portugal because he was one of the grandsons of the late Portuguese king, and his right to the Portuguese crown was at least as defensible as that of any other of its pretenders. Unification ultimately failed, and the result is that in both countries there is today a strange combination of mistrust and, at the same time, nostalgia for what could have been even if few people still want it to be.

Occasionally the dream of a possible unification surfaces on one or both sides of the border, as in the late nineteenth century when the unification of Germany and Italy resurrected in Spain, as well as in Portugal, a faint hope of similar action on the peninsula. Or today when the same hope is cultivated, not in Portugal nor in the whole of Spain but in some nationalistic Catalan minds that see it as a possible solution to their national problem: a Spanish-Portuguese federation that would allow the existence of Catalonia as an equal third partner within the federation. This dream, at least for the time being, remains only a dream.

History has been cruel to the Iberian peninsula; it created two countries where geography seems to have intended only one. And since geography has its laws, history had to create its own counterlaws; that is, both countries had every reason to walk together down the path of history but have insisted on choosing separate ways. Nothing shows this more clearly than their positions with respect to England. While Portugal is the oldest ally of the English, Spain is probably the country that has the oldest claim against them.

More to the point, and equally significant, is the fact that, while the Portuguese dictatorship was acceptable to the NATO powers, the Spanish dictatorship was not. Still, the strategic importance of Spain was such that the United States found it convenient to sign a military agreement with the

Franco regime, trading arms for bases. The peninsula in recent times thus has succeeded in producing both respectable dictatorships and merely acceptable dictatorships. The distinction later had fruitful consequences in the cases of Greece and Turkey.

The two dictatorships pretended to have common international goals and in 1939 signed the Iberian Pact, but the pact had no practical consequences except for reinforcing neutrality during World War II. It certainly did nothing to allay mutual suspicion. That suspicion must be taken into consideration when thinking about the future of Iberia and NATO. Traces of this traditional suspicion already have appeared in the matter of the naval command for the peninsular Atlantic coast. When Prime Minister Felipe González gave as one of the reasons for renegotiation of the terms of Spain's entrance into NATO his reluctance to have a Spanish admiral under the orders of a foreign admiral within Spanish territory, it was understood in Spain that by "foreign admiral" he meant "Portuguese admiral."[1] But it also probably was understood that González's words were the logical consequence of his and his party's nationalistic approach to international politics. The same words, then, could have been applied to any other foreign admiral. Certainly they would have been pronounced, and probably with more gusto, if González had in mind not a Portuguese but an English admiral.[2]

The oldest claim against the English is the third political entity—Gibraltar; it is difficult to call it a colony, much less a country, of the Iberian peninsula. Unlike the other two territories, Gibraltar is a newcomer to the political map of the peninsula but no less of a problem for NATO when thinking of its future in that part of the world. Spain will certainly keep Gibraltar in mind when considering its future within NATO.

Curiously enough, Gibraltar also might become one more problem between Spain and Portugal. In one of the confrontations between Spain and Portugal, the Spaniards occupied a Portuguese border city, Olivenza. Although today it is clearly part of Spain, the Portuguese still claim it as their own; it is not unlike the way the Spaniards claim Gibraltar. The irony is that Olivenza has been in Spanish hands almost one hundred years less than Gibraltar has been an English possession, and that the Spaniards, who so often and so insistently claim Gibraltar, seldom, if ever, even think of Olivenza in spite of the fact that the Portuguese periodically remind them of the problem. Naturally Gibraltar and Olivenza are not problems with identical characteristics, but they do have one thing in common: their inhabitants do not seem to have any interest in changing their present national status. They are also

---

[1]See *Diario de Sesiones del Congreso de Diputados*, 1 December 1982, pp. 50, 71, where the same idea is expressed in about the same words.

[2]The press often comments on possible solutions to the problem of the future command of the Spanish armed forces and always with a nationalistic outlook, implying that few people in Spain are aware of the flexibility of the NATO structure. See, for example, the article in *La Vanguardia*, 19 November 1982; the title in itself is very significant: "OTAN: Canarias bajo mando español" [NATO: The Canary Islands under Spanish Command].

different. Olivenza does not have the strategic importance of Gibraltar and consequently does not have its international resonance; Olivenza does not yet seem to have the popular appeal in Portugal that Gibraltar has in Spain. But the popularity of the Gibraltar issue is a direct consequence of the Spanish government, both dictatorial and democratic. It is just possible that, if Spain ever succeeds in its claim to Gibraltar, the Portuguese may insist on their claim to Olivenza.

NATO AND THE SPANISH ISOLATIONIST TRADITION

Spanish nationalism and its most logical consequence—the traditional neutrality of Spain with respect to European problems—merits additional comment. This neutrality was not a Franco innovation as his defenders pretend. Spain was also neutral during the First World War, and even then it had a long tradition of neutrality, broken only occasionally and mainly with the hope of recovering Gibraltar.

After the Second World War, Spain was isolated not because of its neutrality but, in spite of it, because of its Fascist dictatorship and Nazi sympathies. Naturally the death of Franco, if properly handled, meant that Spain could become respectable again. The need for respectability was strong among Spanish politicians, especially those who, having embraced the democratic faith, still felt tied to the old regime.

This could explain the basic differences of attitude toward NATO among the political parties that dominated Spanish politics before the election of 1983. The two parties that were more closely related to the Franco regime advocated rapid integration into NATO; the parties of the Left, to the contrary, not feeling the need for respectability were ready, and still are, to play the traditional game of isolationism. Ironically, that puts them closer to Franco than the parties of the Right.

Still, the respective positions of all these parties were not altogether clear. The only one that advocated rapid and clear integration into NATO was the Central Democratic Union (UCD). No party felt more pressingly the need for respectability, since many of its members, while in favor of the new regime, had played important roles in the old one. Less convinced, although also in favor of integration, was the Popular Alliance (AP), at the time a very small right-wing party.

The attitude of the parties of the Left was more complex. The Spanish Communist Party was, and still is, clearly against the NATO alliance. This seems to be, in fact, the only point in its foreign policy program. Santiago Carrillo, as well as his present-day successors and their associates, do have a complete and well-articulated foreign policy. But for years they have been giving the impression that keeping Spain out of NATO is an all-consuming obsession. This was made clear again during the last electoral campaign;

whenever a member of the Spanish Communist Party or of its brother party, the Unified Socialist Party of Catalonia, outlined its program, their main point, if not the only one of its foreign policy, was to keep Spain out.[3] At times, trying to appear coherent, they will say that they intended to keep Spain out of both NATO and the Warsaw Pact, pretending to be following traditional Spanish neutrality. But it is obvious, even for them, that neither geographically nor politically is there any possibility of Spain ever joining the Warsaw Pact. Still, it is not very difficult to understand the attitude of the Spanish Communists, and nobody in Spain, Communist or not, can pretend to be deceived by the party's program. This is a theme on which Spanish and Portuguese Communists see eye to eye.

Much more difficult to understand is the attitude of the Spanish Socialists. At the beginning it was similar, if not exactly the same, as that of the Communists. This, incidentally, was not the only point of coincidence between the two parties. Both parties had been persecuted during the long years of the Franco dictatorship, and when they reappeared they were somewhat disconnected from Spanish political reality. Both parties were reluctant to accept a monarchy, which was to be expected in view of their past tradition. But surprisingly, while the Communist Party stressed its democratic faith, the Socialists insisted on being a truly Marxist party, whatever that might mean today. Both parties advocated breaking the continuity between the old and the new regime. However, and this is now history, Spain made a remarkable transition from dictatorship to democracy, and all the political parties had to accommodate themselves to the rapid pace of that transition.

The accommodation was not easy for the Socialists because on some points it meant a Copernican about-face. They naturally were helped by some of their fellow European Socialist parties. For instance, some time before 1980 the British Labour Party sent a delegation to talk to González and give him specific advice: that the Spanish Socialist Party should avoid opposing monarchy since many European Socialist parties could and did get along very well with monarchs; that they should also forget about making Marxism their basic doctrinal credo since the majority of the Socialist parties of Europe had forgotten the subject a long time ago; and finally, that the Spanish Socialist Party should downplay its opposition to NATO.[4]

There does not seem to be any public evidence of similar efforts having reached González from other European Socialist parties, but there can be little doubt that such advice was given. Very soon, even before the king took his famous stand against the military rebels in 1981, the Socialists accepted the

[3]The main leader of the Catalan Communists, Lopez Raimundo, during the last electoral campaign went so far as to say that the mission of the Communists was to remind the Socialists that Spain should abandon NATO. *La Vanguardia*, 8 October 1982.

[4]The information about this delegation and its message was given to the author of this paper by a prominent member of the British Labour Party, Sir Geoffrey de Freitas, on 13 September 1980 in Norwich during the course of a colloquium organized by the Seminar in Atlantic Studies, University of East Anglia.

monarchy; and although it cannot be said that they completely abandoned Marxism, they never talked about it after the pro-Marxist faction lost the ideological battle during the party's 28th Congress.

THE REFERENDUM QUESTION

The change in Socialist policy toward NATO was less spectacular but not less real. When the Spanish government proposed joining the NATO alliance, the Socialists demanded that the matter be put to referendum and be decided by the Spanish people. But why a referendum for this, and only this, question? It is tempting to think that it is another version of what Americans call "passing the buck." This might have been the case when it first was proposed; it was an easy way of postponing the issue when the UCD government was ready to sign Spain's incorporation into NATO. It was also a way of not antagonizing the Communists, which at the time was very important for the Socialists since they did not expect to win sufficient votes to govern without Communist help.

If that was their reasoning, they soon abandoned it for a more sophisticated argument: joining NATO should be part of a global negotiation in which Gibraltar and the Spanish entrance into the European Communities also should be discussed. This is, in fact, what Ernest Lluch said in a Catalan radio interview a few days before the last election. Lluch, at the time the party's spokesman in parliament and subsequently the minister of health and therefore a qualified voice within the party, made it very clear that the party was not against Spanish membership in NATO. The problem was not of substance but of form: the UCD government had joined NATO without getting anything in exchange, giving away one of the few cards Spain could play in the game of international politics.[5]

Asked about the referendum he naturally reaffirmed the party's commitment to it, without being too specific. Fortunately on this matter there is better and fresher information given by a voice still more qualified than Lluch's. González himself, answering a question in the Spanish parliament, made it clear that it was not his intention to introduce the referendum too quickly, and that when the proper moment arrived it will have to be preceded by a campaign of information so that the Spanish people may vote with sufficient knowledge on the matter.[6]

---

[5]The interview was aired by *Radio Nacional de España en Barcelona* in one of its Catalan language programs, 18 November 1982.

[6]González's statement was made in parliament, 1 December 1982. Since then the same idea has been repeated not only by him but also by numerous Socialist leaders including: Alfonso Guerra, vice-president of the government, during an interview given to Fernando Jauregui and published in *El País*, 5 February 1983, and again in another interview, this time on the first channel of Spanish television, 11 April 1983, and reported in *La Vanguardia*, 12 April 1983;

Since González made that peculiar statement, which at least by his enemies could be interpreted as an intention of manipulating the vote, the Socialist government has had the responsibility of governing a country with enormous economic problems and with the highest unemployment rate in Europe. Issues much more pressing than the more doctrinal one of Spain's status in NATO have had to be confronted. As is well known, Spain is now a member of the alliance, but its armed forces are not integrated into its command.

This situation, similar to that of France, would perhaps have been the happy compromise that could satisfy everyone but for the Socialist preelectoral promise of holding the referendum, a promise that keeps haunting the Socialist government. Specifically, the leaders of the Communist Party and of its affiliated union, Comisiones Obreras, take advantage of every possible opportunity to press the government on the matter of the referendum.[7] Even within the Socialist Party, now and then voices ask for the referendum. This is also the official stand of the Socialist union, Unión General de Trabajadores, which adopted a resolution asking the government to keep its promise to hold a referendum "as soon as possible."[8]

It is even tempting to suspect that the government itself is somewhat divided on the matter. While González has made it clear that there will be no referendum for the time being, other Socialist leaders have repeated the old commitment to hold the referendum soon. González's explanation for the indefinite postponement seemed to be a definite solution to the problem: since Spain is now a faithful ally of NATO, it is not going to do anything that could weaken the alliance as long as there are tensions between East and West.[9] That is, for as long as the United States and the Soviet Union fail to reach a meaningful accommodation, Spain not only will not abandon NATO but also will not even hold a referendum on the matter. This is certainly the position of a statesman. But González the statesman is also a politician, and, since the statesman cannot contradict himself, his second in command reaffirms the commitment of the party to the referendum, as does the minister of

---

Fernando Morán, minister of foreign affairs, in an interview given to Miguel Angel Aguilar and published in *El País,* 6 March 1983; Saez Cosculluela, leader of the Socialist parliamentary group, in an interview given to Luis Saez and published in *Informaciónes*, 7 February 1983; and Narcis Serra, minister of defense, in declarations made in Brussels and reported in *La Vanguardia*, 3 June 1983.

[7]At times well-known Socialists appear in anti-NATO marches; for instance, Pablo Castellanos, president of the Comisión de Justicia of the Spanish parliament, led the third march of anti-NATO protest held on 20 March 1983, as reported in *La Vanguardia*, 21 March.

[8]This specific resolution, adopted during the thirty-third general meeting of the union, held from 1 to 5 June 1983, was somewhat marginal and perhaps is a concession to the left wing of the union, as suggested by the *ABC* editorial comment of 9 June 1983. The resolution appears also in *ABC*, 6 June, apparently the only major newspaper that reported it.

[9]The statement was made during a press conference held on 14 March 1983 and reproduced by practically all Spanish newspapers on the following day. See, for instance, *La Vanguardia* and *El País*, 15 March.

foreign affairs. The situation is then somewhat confusing. Spain is temporarily halfway within NATO until some unscheduled event in the future—the referendum—decides to become fully integrated in it, or definitely to abandon the alliance, or to maintain the status quo.

The observation that Spain is temporarily in NATO may bring a cynical thought to mind: Spaniards like to say that in their country nothing lasts longer than a temporary situation. The phrase cannot be applied here since the present Socialist government is not a temporary government; it is a definitive one, although for only four years. Sooner or later within this period of time the government will have to face the issue. The reasons that prompted the Socialist Party, first to demand a referendum from UCD and then to promise that if elected it would hold one, are still there and will increase in importance as the next election approaches. But more than anything else, they will not disappear because the government has a clear commitment to the people who voted for it.

Given González and his colleagues' moralistic approach to politics, they will stand by their commitment. But no drastic measures are foreseeable. González, on this as on so many other subjects, has become the personification of prudence. Certainly the present Spanish government is not going to accelerate the process of incorporating the Spanish armed forces into the NATO command structure, but neither is it going to withdraw Spain from the alliance. All indications are that the next two or three years will be spent searching for a formula that, without changing the country's commitment to NATO, can be offered to and accepted by the Spaniards. The finding of this magic formula, though, is not solely the responsibility of the Spanish government but also of the other NATO powers. The strategic importance of Spain for the defense of the West, and more specifically of the western Mediterranean, is so clear that it is unnecessary to elaborate on it.

It would be a mistake to think that a future change of government also will imply a change in the Spanish attitudes toward NATO. Given the forces that today dominate the political map of Spain, it is foreseeable that after the present Socialist government Spain will have either four more years of the same or four years of the AP, which became practically the only opposition party in the last election. The AP is not going to be less nationalistic than the Socialist Party, and in certain matters, such as on the subject of Gibraltar and on the similar one of Ceuta and Melilla, it may become more demanding than the Socialist. After all, Spanish neutrality is a tradition of the Right that only recently has been taken up by the Left.

FRANCE, ITALY, AND NORTH AFRICA

France and the Mediterranean*

ANDRÉ MARTEL and CLAUDE CARLIER

The special nature of France's participation in the defense of NATO's southern flank results as much from its historic experience as a Mediterranean power as from the specific character of its unique role within the Atlantic Alliance. Thus the Mediterranean world is of primary significance for the function it fulfills in France's conception of North-South relations; it is, in that sense, a crucially important geopolitical area. Its strategic role in which episodes of the East-West confrontation are played out is secondary to that primary function.

Along with Spain, France is the only European member of the alliance to have both a Mediterranean and an Atlantic shore. Over the ages, the Mediterranean had forged French mental structures, well before the Atlantic imposed another dimension of thought and action. This latter process lasted from the turn of the sixteenth century until the advent of the Fifth Republic. But the orientation of NATO toward the interests of the North Sea countries and the Anglo-Saxon bloc disturbed de Gaulle's France and revived French interest in its Mediterranean connections.

THE HISTORICAL CONTEXT: A GEOPOLITICAL APPROACH

Does the heritage of ancient history deserve to be taken into account? Should scholars preoccupy themselves with galleys in the era of nuclear submarines? Those who sail the submarines and those who develop them do belong to the twentieth century and are preparing the way for the twenty-first; they are, in spite of all that, no less the product of a past which conditioned them.

Until the end of the fifteenth century, the Atlantic Ocean played the role of a natural barrier for the languages, beliefs, and merchandise as well as for the peoples of Africa, Asia, and Europe who were living along the Mediterranean littoral. They could only leave through the Strait of Gibraltar where they would run into the oceanic unknown or filter through the often hostile land routes of the bordering countries. All of those routes did not offer the same ease. To the north, the Don, Dnieper, Danube, Vardar, Po, Rhône,

*Translated by Mark R. Rubin.

and Ebro drain temperate countries that are favorable to human activity. To the east and south, with the exception of the Nile, the Mediterranean fringes are isolated by immense deserts. Some of the caravan routes that cross those regions lead to black Africa; others lead to Mesopotamia, Persia, India, and China, the lands of the oldest civilizations.

It was from Asia that the essential forces came which formed the present-day Mediterranean, but those contributions did not correct the natural lack of balance between the north and south shores of the sea. Instead, they aggravated it even though political or religious unification occasionally scored durable victories. In the north, the Indo-Aryan heritage was divided into two linguistic areas—Greek and Latin—subsequently perpetuated first by the Roman Empire and then by Christianity. The Western Empire was profoundly changed by the Germanic invasions, while the Eastern Empire experienced waves of Slavic migrations. In the east and south, the survival of Christianity was riding on the Arab-Moslem conquest which, stopped by the ocean, poured into black Africa. The holy wars, *jihad* and crusade, were the daily reality of seven centuries of history, at the end of which Christianity and Islam appear to have stabilized their front, if not their frontiers, at the Strait of Gibraltar and at Sicily but continued to dispute the lands around Constantinople.

The kingdom of France took part physically and spiritually in the confrontation that founded the Carolingian dynasty and gave its patron saint to the Capetians: Louis IX, the archetype of the crusader-king. But before him French jurists had proclaimed to one and all that their master, the very Christian king, owed his crown to God alone. That was a way of pointing out, in an evangelical vision of the separation of the spiritual and the secular unknown to Islam, the sovereignty of the king of France, as compared to the Holy Roman Emperor or the pope.

Upon the linguistic differences and religious wars are superimposed the national conflicts, whose legacies constitute most of the roots of contemporary conflict. The Middle Ages were ended, in fact, by the Ottoman conquest of Constantinople in 1453, but the "Eastern question" which arose from that event is only one of the aspects of the struggle for possession of the world provoked by the discovery of America forty years later. While the Turks of central Asia were seizing the old Byzantine Empire and imposing Moslem domination as far as Hungary, the Christian Spanish and Portuguese were opening up the Americas, black Africa, and Southeast Asia to European settlement, exploitation, and religious conversion.

Europe, thus projecting itself overseas, was nevertheless no longer synonymous with Christendom. Nation-states developed and affirmed their existence by fighting with one another. In order to survive, France was first to accept the Sublime Porte against the house of Hapsburg, which was threatening to establish a universal empire. And in the Barbary and the Levant states, French consuls settled. They fancied themselves to be the protectors of all Latins, including Christians who had been able to practice their faith in the East under the domination of the Moslems, then Arabs, and finally Turks.

This alliance of opposites (continental and Mediterranean) was intended to cover the land borders of France, while it was playing out on the oceans the sort of decisive expedition in which victory would allow the independent French nation-state to become a major power. The war against England lasted nearly two centuries and ended with the elimination of France from the American continent and from India, in spite of the belated effort to ally French and Spanish destinies to prevent the formation of a British maritime empire.

In order to reopen the Indian question, the new French Republic attempted to establish itself in Egypt. The Turks could only turn toward the English, all the more determined to support Ottoman integrity since the Russian push toward the Straits, the Caucasia, and the Caspian Sea. This expansion had attracted the interest of the tsar, the master of the one true continental empire, who reigned from the Vistula to the Sea of Okhotsk. England won Malta, Cyprus, and Egypt without France having to resign itself to disappearing from the Arab-Moslem world. The breakup of the Ottoman Empire in the nineteenth century provided France with the occasion to reappear in force. It established its dominion over Algeria, Tunisia, Morocco, and at the same time on the southern rim of the Sahara which also was populated by Moslems.

This second phase of worldwide colonization provoked, much as did the first one, rivalries among European competitors. But France, because of the action of compensations on a global scale, benefited from the support of Germany and England against the claims of Spain and Italy, the latter two having been obliged to be satisfied with Morocco on the one hand and with Libya on the other. While the Italian and Spanish conquerors of the nineteenth century evoked self-images of Rome and the Crusaders, the divided and demoralized Arabs of the Mediterranean littoral could compensate with the memories of Andalusian Splendor and the successes of the *jihad* in North Africa.

Nevertheless, it was to France and England that the eastern Arabs turned during the First World War to shake off the Ottoman yoke. Their coreligionists in the western Arab lands counted in vain upon a German and Turkish victory to rid themselves of the French, Spanish, and Italians. The victorious powers of the Entente reaffirmed their domination over the Maghreb and established their mandate in the name of the League of Nations. France moved into Syria and Lebanon in its role as protector of Christians, while England played the card of Arab national emancipation in Iraq, Palestine, and Egypt. France itself feared the consequences of an ultimate liberalization in its Levant "mandates" or in its Algerian colony and other protectorates in the Maghreb. Arab nationalism, stimulated by European models, could challenge their control. An even more disruptive element was the building of a Jewish national homeland under the British mandate which energized Zionist nationalism to compete with the Arab presence in Palestine.

It is against that background of transmitted or learned experience that generations since the end of the First World War either have witnessed or participated in events of worldwide importance. Those events may be seen as occurring on two inseparable mixed levels: decolonization and the Cold War.

Decolonization in the Mediterranean

France had difficulty perceiving the irreversible nature of worldwide decolonization, the roots of which during the period between the two world wars had an especially profound effect on the two poles of its empire—Indochina and North Africa. For France, it seemed not so much a question of the fundamental overthrow of its political domination, which had so long existed, as the work of local agitators supported by its enemies and allies, both of whom were equally jealous of its power.

It was an era when three streams of thought reanimated resistance by the conquered peoples. The ideas of liberty and equality brought by French schoolteachers and missionaries ran counter to the colonial system, while the Wilsonian principles proclaimed by the League of Nations justified a peoples' right of self-determination. At the same time, the Communist Third International called for the exploited peoples of the East to break the colonial yoke. Nevertheless, in the Arab Mediterranean world the liberals and Communists were rapidly overtaken by religious militants who drew from the reawakening of the Moslem world, and especially the Arab *Nahda*, their strength and will to give to their emancipation the character of a return to an idealized past. This religious nationalism disturbed French liberals, the Socialists in particular, who were more concerned about progress and assimilation than about historical turnabouts.

Four defeats shook French domination of its colonial possessions. That of 1940 in France provided colonial peoples the hope of a German victory. That of 1941 in the Levant consecrated British supremacy in the Near East. Then in North Africa itself France backed down in the face of demonstrated American superiority in 1942. France's participation in the victorious campaigns of 1943–45 did not wipe away the certainty that thereafter the former colonialists would no longer be masters of the world. Finally, the defeat of 1954 in Indochina, followed immediately by the Bandung Conference, revealed the worldwide breadth of the Third World's demands.

Even before the existence of the United Nations—where later American and Soviet anticolonialism were mutually supportive out of a common concern with eliminating the European empires—at the end of the Second World War, the Arab League took up the task of liberating those Arab nations that were still colonies. In Lebanon and Syria, French forces were eliminated in 1945 due to the active support of the British. Libya became independent in 1951, thus escaping the concerted projects of England, France, Italy, and even the Soviet Union. Tunisia and Morocco became sovereign states in 1955 without any actual fighting.

The Algerian war, on the other hand, lasted until 1962 and caused grave tensions in France between liberals and partisans of a French Algeria as well as between Gaullists and members of the Secret Army Organization. The result of French withdrawal from Algeria was the return to France, in a period of economic expansion, of more than 1 million *pieds noirs* (Algerians of French or European origin). For them, Tunisia, Morocco, and Algeria remained their "lost land."

The Cold War

The Cold War did not result from decolonization any more than it blended in with it. Since the creation of the Third International, and through the intermediary of national Communist parties, Soviet foreign policy interests have fed on the problem of decolonization without having engendered it and have used it without mastering it. German bolshevism during the Weimar Republic sought to weaken France by calling for strikes by workers, insubordination by soldiers, and uprisings by colonists. The French Communist Party openly supported the Riff and Syrian insurgents. The tactical about-turn of the Third International, beginning in 1935, marked treachery that was never repaired: the Soviets for a time stopped supporting colonial demands for independence in order not to weaken bourgeois France in the face of the new threat of Nazi Germany.

The advent of nazism crystallized tendencies in France in a totally different manner. In order to stand against Germanic revisionism, which blended with totalitarian paganism, France resigned itself to an alliance with the Soviet Union. But the Spanish Civil War caused virtually the opposite to emerge. Both conservatives and moderates, who had stifled fascism in France, had considered the possibility of a Mediterranean bloc which would bring together countries that had adopted military regimes like the Greek or Yugoslav monarchies, or the avowedly Fascist government of Italy, to stop international Communist revolution. In a way, Spanish Francoism, which was working to overthrow the Popular Front, seduced them in a lasting manner. On the other hand, the French left, hostile to all dictatorships and friendly to all refugees, felt close to the Spanish Republicans. Nevertheless, all moderate factions believed that the British alliance and a guarantee of eventual support from the United States was the only way of standing up to Germany.

The Second World War resolved the dilemma but not without internal strife. The Soviet Union, after being aligned for a time with Germany, became the ally of the Anglo-Saxons. The Communists took an active role in the Resistance. Out of fear of Germany and an Anglo-American hegemony, de Gaulle allied himself in 1944 with the Soviet Union. Faithful to the policy of European equilibrium, the Fourth Republic nonetheless concluded the treaty at Dunkerque, which linked France and England and paved the way for an extension of the treaty to Benelux. Soviet imperialism ultimately led France to desire an Atlantic Alliance and to agree to the integration of its forces under the nuclear protection of the United States.

Why in 1966, but with roots as far back as 1952, did France denounce any scheme of military integration and retain only the form of an alliance? Among the different reasons cited, the only ones that are important to this study are those that relate to the Mediterranean. It was there in 1956 during the Suez crisis that France realized the extent of its dependence. The Americans, who still controlled the United Nations, did not want to risk a confrontation with the USSR, which had successfully deceived the West into believing that it had attained nuclear equality, nor did they want to risk the condemnation of the newly emerged Third World which was so skilled in denouncing neocolonialism. De Gaulle from his retirement, Pierre Mendès-France as early as 1954, and Guy Mollet in 1956 all concluded that it was necessary to accelerate the process of constituting a French nuclear force that would be the guarantee of freedom of action.

The nuclear heritage of the Fourth Republic facilitated de Gaulle's Fifth Republic ambition to construct a sovereign nuclear deterrent. Nevertheless, if foreign policy independence for France was an end, it also appeared to be the principal condition of a new program of influence that developed two contexts: the North-South dialogue and a diplomatic realignment. The Mediterranean countries were situated in the first of the concentric circles of France's ambitions.

The North-South Dialogue

In the Gaullist version of worldwide equilibrium, Europe was to become a third pole of power between the United States and the Soviet Union. It could not succeed in that endeavor unless Europe found in the Third World the territory and energy that it lacked. De Gaulle's successors reduced that vision to a desire for at least equilibrium between the United States and Europe within the Atlantic Alliance. The petroleum crises that began in 1973 made an agreement with the Arab countries indispensable. The success of that new French Third World policy presupposed that various obstacles would be removed: the privileged alliance with Israel, the dissension among the Arab states, and the fears of black Africa.

Israel, whose postwar settlement France had been aiding from its Mediterranean ports (Sète in particular) and whose victorious defense it applauded in 1948, became an ally in 1956 during the Suez crisis. The ties created at that time were strengthened by an emigration from the Maghreb which created a French-speaking community in Israel and reinforced the French Jewish community through the addition of Sephardim. However, de Gaulle had no intention of subordinating his Arab policy to Israeli vetoes and seized the occasion of the 1967 war to distance himself from a country that was too close to the United States. The 1968 retaliatory raid on the Beirut airport, following the hijacking of an El Al plane, allowed him to appear as the protector of Lebanon and to redirect French sales of *Mirages* away from Israel. After him, French heads of state maintained the same diplomatic line of de jure recognition of Israel's existence within its pre-1967 borders.

In North Africa the pro-American shift in Egyptian policy after the 1973 Yom Kippur War reopened historic Arab cleavages. Morocco and Tunisia aligned themselves with moderate Arab countries supported by the United

States. Libya and Algeria, both petroleum producers, put themselves into the camp of the intransigents for whom American-Zionist imperialism remained the ultimate evil. To the great chagrin of its former protectorates, France accorded Algeria favored treatment, explained by its desire to prove the historical and political merits of its disengagement as much as by the politics of energy. The same was true of Libya, whose petroleum wealth also made it an esteemed client even if the Arab-Islamic ambitions of its leader exploited latent conflicts.

Between Algeria and Morocco, the Cherifian expansion provoked the Polisario crisis. At first, France supported Morocco and Mauritania but then withdrew to a position of supporting the status quo. In order to resist the Polisario adversary benefiting from Algerian diplomatic protection and Soviet armaments, Morocco accepted an increase in American aid. In Tunisia, on the other hand, France logistically supported the army in suppressing an uprising inspired by Libya. To the south of the Sahara, French policy makers showed more hesitation in facing the Arab-Moslem imperialism of Colonel Qaddafi, who sought to destabilize central Africa, Chad, and indeed Niger, which was valuable for its uranium reserves. Still, France remained faithful to its agreement for military cooperation, which pointed out the difficulty of reconciling an Arab policy with an African policy.

A Mediterranean Community

De Gaulle had sought to create a Europe led by France and founded on Franco-German understanding reinforced by its African extensions. His successors, while not abandoning the Paris-Bonn axis, have given more importance to France's Mediterranean partners. Among other things, they realized the need to counterbalance the affinities that united the Anglo-Saxons, Germans, and Scandinavians.

Various French political tendencies lent themselves to this operation but with different ambitions. Those who were anxious for a "new order" had placed their hope in Salazarism, Francoism, neofascism, and even in the Greek colonels. The possibility of a Eurocommunism, inspired by the Italian and Spanish models, appealed to others who could not resign themselves to the archaism of the French and Portuguese Communist parties. A neomonarchic liberalism had brought Paris and Madrid together. Social democracy had waited a long time to come to power in France, Spain, and Greece. It is, however, important to note that all had sought, and continue to seek, links to the states of North Africa and the eastern Mediterranean: military Turkey, pro-Soviet although anti-Communist Libya and Algeria, Cherifian Morocco, and Israel under the Labour bloc.

It is true that the existence of 4,200,000 emigrants, of whom more than 3,000,000 are Mediterraneans, encouraged this orientation. Among the latter, Italians (450,000), Spaniards (410,000), and Portuguese (810,000) remain as potential French citizens. As opposed to the past, Spaniards and Portuguese now think less about settling permanently in France than about eventually returning to their own country. Moslems appear to have more difficulty assimilating, to the extent that their countries of origin, especially Algeria, wish to protect their religious, cultural, and national identity. The presence of 820,000 Algerians, 440,000 Moroccans, 200,000 Tunisians, 120,000 Turks, and 120,000 Africans poses grave political and social problems within France.

Those problems, however, should not be confused with the much more serious ones caused by the activism of several groups of emigrants who harbor or are actively engaged in international or regional terrorism. The prudent restraint with which French governments have treated Arab and Israeli terrorists or counterterrorists, "Red Brigadists," or Basque nationalists has caused irritation in neighboring capitals. On the other side of the issue, it is appropriate to note the opening of French military schools to Arab and African officers who are designated for important commands or high administrative or technical positions. In 1981, out of 3,565 foreign trainees, 1,120 came from Arab countries and 1,980 from black Africa. Through this program France expects to maintain or to increase its cultural, political, and commercial influence, especially in the area of arms sales.

In reality, it is the economy that constitutes the true obstacle to a Mediterranean community. How is it possible to create, on a linguistic or cultural basis, an association of countries which are equally destitute of energy resources and that are agricultural competitors? Even within the Common Market, France and Italy confront each other on the wine issue. Spain would like to enter the European Economic Community not only to sell its wine more easily but also the kinds of fruits and vegetables that France and Italy already have difficulty selling in their own markets or in northern Europe. This competition complicates even more the desire of the European Mediterranean powers to diversify their industrial exports to the countries of the Maghreb. Libyan and Algerian petroleum and Algerian gas sales bring France a part of the energy resources that it lacks and give its Arab trading partners some buying power, notably for the purchase of arms. But those contracts are sometimes thought to impose financial and political constraints that only add to the dependence created by the Soviet gas deal.

FRANCE AND THE MEDITERRANEAN GEOSTRATEGIC ARENA

The southern area is for France only one of the elements of a larger whole. It must be used to cope with the global danger that brought France into the Atlantic Alliance in the first place and which justifies its continuing membership today. In this context the Mediterranean is first a "maritime" sector connecting with a second "continental" sector, both of which also border on the Atlantic. The French evaluation of a possible theater of operations in the Mediterranean does not depend only on its permanent geographic features. Its specificity arises from the relationship between the technical characteristics of the moment and from the parameters of the environment.

The Mediterranean Environment

Even in the age of nuclear submarines, supersonic aircraft, and radar, the Mediterranean remains very much as it is represented on a map and by the map scale. This inland sea opens on the Atlantic through the Strait of Gibraltar and communicates with the Indian Ocean by way of the Red Sea. It may be divided into three basins from west to east: the western Mediterranean, the eastern Mediterranean, and the Aegean. The coastlines, islands, and peninsulas that delimit and subdivide them also facilitate their surveillance and defense. Gibraltar, Malta, Suez, and the Turkish Straits mark these divisions both physically and symbolically. For want of many estuaries and tides, large sheltered deep ports are rare. Nevertheless, the rocky coastlines offer numerous refuges against the brutal wind.

The powers that border on the Mediterranean have at their disposal national systems of naval and air bases, resulting from the superimposition of new technologies on sites already predestined for defense use by their physical characteristics and by their access to the interior. These large civilian or military complexes have lost less of their importance than many planners had anticipated. In spite of improved cruising and operating ranges, both ships and airplanes still require supply bases and can only be certain of their accessibility if they are on their own territory or, more problematically, within the framework of alliances.

French Bases

As illustrated by publicly available data, French forces in the Mediterranean are attached to a network of ports, from the Pyrenees to the Alps, and in Corsica as well, which has been there since the days when the navy was composed of sailing ships. Attracted by economic activity and population density, airports have sprung up in the same places. Marseille and Toulon, supported by Sète and Nice, constitute the principal naval and air force facilities. At the same time, the lower Rhône Valley has lent itself to the construction of airports at Marignagne for international civil traffic and at Orange and Istres for military aviation. The port of Nice also was reinforced by an international airport. Nîmes offers its air facilities between Sète and Marseille. The expanded network of Air Inter airports and the growth of civilian marinas, coupled with old fishing ports, further multiply the number of shelters open to small units or formations. These complexes are often complemented by industrial zones, naval shipyards, or refineries.

Crises in the Mediterranean are likely to require refueling and replenishing at some distance from French territory. Plans must be guided by the existence of foreign bases, either friendly, neutral, or adversary. Certainly the air and sea space controlled by the two Mediterranean partners of France—Spain and Italy—surround the western Mediterranean, providing excellent facilities for surveillance and interception. However, Morocco, Algeria, Tunisia, Libya, and Malta, which lie on the other side of the Mediterranean, form an officially neutral barrier but one that in reality is more or less open to a possible adversary depending on the policies of the moment. Contingency planning for a probable crisis or conflict cannot ignore the former French and English bases of Mers-el-Kebir, Bizerte, and Malta.

French Power

France has always dedicated the essential part of its military defense to its northern and eastern borders where it faced continental enemies and has divided its naval forces between the west and the east, sometimes giving priority to the Atlantic, sometimes to the Mediterranean. Its present doctrine has changed nothing in that priority, even if it takes into account the fact that the potential enemy is no longer camped at its borders and even if it bases its defense on the ability to make deterrent strikes on the adversary's cities. Nevertheless, the general evolution of its means and the composition of its forces illustrates the importance given progressively to its southern flank.

French armed forces, under the authority of the head of state, are organized into three armies responsible for recruitment, training, management, and operations. During peacetime, military, naval, or air regions maintain those forces. In time of war, they may be assigned, either wholly or partially, to theater commands.

In the Mediterranean area the Fifth Military Region (Lyons) covers the same territory as the Fourth Air Region (Aix-en-Provence), and its shoreline corresponds to the one under the jurisdiction of the Maritime Prefect of Toulon. Roussillon, Languedoc, Provence, the French Riviera, the Alps, the Rhône Valley, and Auvergne all make up the deep inland littoral of the coastline extending from Port-Vendres to Menton and taking in Corsica, which also is included in the Fifth Military Region.

This structure has the advantage of a logical Mediterranean defense scheme and depends upon an Italo-Alpine defense concept. On the other hand, it does not rest on French-Spanish cooperation across the Pyrenees. This peacetime configuration reflects the double Atlantic and Mediterranean coherence to the rear of the French battle groups stationed on the Rhine and does not in any manner prevent the creation of operational commands as dictated by the situation.

French Forces

The nature, size, and disposition of the forces in the naval, air, and military regions of southern France demonstrate the priorities of the Fifth Republic. Placed under the command of the Strategic Air Forces and acting on the direct orders of the president of the republic, the launch centers for surface-to-surface nuclear missiles are buried under the plateau of Albion not far from the *Mirage IV* bases in Istres and Orange. Recent visits and declarations by the president, prime minister, and minister of defense indicate the government's determination to harden the present missile silos before adopting a system of mobile launchers.

It was by the withdrawal of its Mediterranean naval forces in 1959 that France first underlined its intention, either to effect a substantial increase in French influence within NATO or to disengage itself from the military portion of the alliance. In 1974 the Defense Council reassigned two renovated aircraft carriers, the *Clémenceau* and the *Foch*, to the Mediterranean. The squadron then rearranged itself around those two ships, surrounding each with a cruiser and guided-missile frigates. The force at sea is reinforced from the land by naval air forces stationed at Hyères, Canjuers, and Nîmes. The squadron also maintains an amphibious capability to land mechanized forces. A submarine flotilla, which since January 1983 has included a nuclear-powered attack boat, completes the Mediterranean squadron.

Air forces may appear weak when compared to naval forces, but the existing infrastructure shows that impression to be incorrect. Military aviation bases (naval, air, and army), as well as civil airports, can quickly receive aircraft of the Tactical Force, whose command is coupled with that of the First Air Region. Thus, it is possible to envisage both preventive sorties by aviation units that are on alert and reinforcement operations from more northern bases. The network is completed by radar stations at Nice and Narbonne.

As for the army, its territorial forces, without being negligible, are less important for strategic plans than the foreign or French elements that form the Rapid Assistance Force. The latter is not completely stationed in the Fifth Region, but various airborne or mechanized regiments are garrisoned in either Carcassonne or Orange. For the last year the Thirty-first Brigade, whose home base is in Aubagne near Marseille, has constituted its principal nucleus. Future attachment of the Twenty-seventh Alpine Division to the Rapid Assistance Force would complete the goal of making an operational reserve capable of acting either together or in separate units on the Rhine or in Friuli as well as in Tunisia or Chad, as long as there are means of air transport.

While the missions of those southern forces are basically the same as those of French defense forces in their totality, they are adapted to the particular conditions of a maritime environment. Through its three elements, one of which is permanently installed in Provence, the Nuclear Strategic Force assures the deterrence upon which French defense policy is founded. It is the task of the other forces to provide the cover necessary to ensure its credibility. Surveillance of coastlines and air-approach routes is of more critical importance in the south than is monitoring of the land borders. Special mention should be made of the practice "war" against submarines sent to test the defense apparatus. The airborne and land-based naval forces are also available to fulfill the contractual obligations of the French government, either within the Atlantic Alliance or to the benefit of its different African allies as well as at the request of the United Nations or other international organizations.

The deployment of the Mediterranean squadron indicates that France does not intend to let the superpowers maintain a naval power monopoly there. In an area through which the essential portion of its energy supplies passes, France will continue to keep credible naval, air, and military forces.

CONCLUSION

What has been the contribution of the Mitterrand government which has been in power in France since May 1981? The most recent presidential and ministerial declarations on defense suggest continuity with the policies of the recent past. French defense remains independent and is armed with the nuclear and classic means of staying that way. The hardening of missile silos and the creation of a multipurpose intervention force are only the most spectacular manifestations of that independence. Ironically, the capabilities promoted by the policy of independence of French military and naval forces in the Mediterranean, promoted by various administrations of the Fifth Republic, permit an informal but close collaboration that strengthens both France and NATO in the area.

U.S. Security and Italy: The Extension of NATO to the Mediterranean, 1945–49*

E. TIMOTHY SMITH

Although originally intended as a security system encompassing the North Atlantic, the North Atlantic Treaty Organization has since become a North Atlantic-North Mediterranean alliance. A key nation in this development was Italy, which joined as a charter member in April 1949.[1] Seen as both a Western European and a Mediterranean country, Italy's strategic location and Western interest that it remain a non-Communist nation led to its incorporation in the Brussels Pact. As early as 1945, American officials expressed concern over Italy's relationship to the West because of its strategic position in the Mediterranean. As the Cold War developed and the U.S. perceived the Italian Communist Party (PCI) as bent upon taking Italy into the Soviet bloc, the United States became worried about the loss of that strategic nation through internal subversion. For that reason, and because of French interest in protecting its northern African territories and the security of the Po Valley, Italy became the first Mediterranean member of the North Atlantic Alliance.

THE SECURITY ISSUE, 1945-48

During and after the Second World War the United States emerged as the dominant external power in the Mediterranean region and in Italy. The United States came to play a major role there for two main reasons: Italy's economic condition dictated a long-term dependence on U.S. aid and the U.S. concern over the possibility of PCI domination of that strategic location.[2] The PCI program immediately after the war sought collaboration with the

*Portions of this paper appeared in *Diplomatic History* 7, no. 2 (Spring 1983): 139–55, as "The Fear of Subversion: The United States and the Inclusion of Italy in the North Atlantic Treaty."

[1]Stephen E. Ambrose, in *Rise to Globalism: American Foreign Policy, 1938–1980* (New York: Penguin, 1980), p. 156, argues that "the presence of Italy . . . among the members of the alliance made a mockery of the words 'North Atlantic' in the title. . . ."

[2]Muriel Grindrod, *The Rebuilding of Italy: Politics and Economics 1945–1955* (London: Oxford University Press, 1955), p. 129.

Italian center parties. The Communists participated in the government until 1947 and assisted in drafting the Italian constitution. Stefano Silvestri has written that the political legitimacy this gave them has prevented the kind of division in Italy that appeared in Greece after the war; the PCI therefore has been seen in Italy "more as an estranged brother than as a complete enemy."[3]

Despite the moderate public goals of the PCI, U.S. officials perceived the party as a potential revolutionary force and as a tool of Soviet foreign policy. It was seen as a threat to U.S. strategic, political, and economic objectives in Italy. American unrest over Italy's security emerged in the discussions on the peace treaty with Italy. The Ad Hoc Committee of the State-War-Navy Coordinating Committee expressed uneasiness that certain "forces" threatened to sweep Italy into an international political alignment diametrically opposed to U.S. interests. Of greatest concern to the committee was Italy's strategic location in the Mediterranean and American dependence on Mediterranean sea lines of communication to the oil supplies of the Near East.[4]

Two key events heightened U.S. interest in the security of Italy: the 1947 Greek and Turkish situations and the 1948 Italian election. These events reiterated to U.S. policymakers Italy's significance as a pivotal nation between Mediterranean interests to the east and European interests to the north and west. The Greek and Turkish problems led to a fear that communism would spread west through Italy into France. The nervous interest in the Italian election of 1948 was related to the possibility of a PCI victory. If the PCI won that election, many believed that a strategic location would be lost. Soviet influence and communism could then spread with much greater ease to Greece and France.

The civil war in Greece and the Soviet pressures on Turkey, which prompted the Truman Doctrine in March 1947, led to numerous warnings about the danger of a domino effect: the spread of communism west out of Greece. Policymakers in the United States believed that, if either Greece or Turkey became Communist nations, there would be profoundly disturbing psychological and political reactions in all nations of that region. American officials feared that stability in that vital area of the world would be disrupted, and the faith of nations in the ability of democracy "to maintain itself" would be weakened.[5] The Truman administration's policy of aid to Greece and Turkey seemed to halt that perceived Soviet thrust into the eastern Mediterranean and deepened the U.S. commitment in Europe.

---

[3]Stefano Silvestri, "The Left and Security Problems in Italy, France, and Spain," in *The European Left: Italy, France, and Spain,* ed. William E. Griffith (Lexington, MA: Lexington Books, 1979), pp. 146–47.

[4]Report by the Ad Hoc Committee of the State-War-Navy Coordinating Committee, 6 September 1945, U.S., Department of State, *Papers Relating to the Foreign Relations of the United States, 1945* (Washington: Government Printing Office, 1968), 4:1037–40. (Hereafter cited as *FRUS* followed by the appropriate year).

[5]U.S., Congress, House, Committee on Foreign Affairs, *Assistance to Greece and Turkey: Hearings Before the Committee on Foreign Affairs H.R. 2616,* 80th Cong., 1st sess., 1947, pp. 385–86.

From that point on Italy became the center of American concern in the Mediterranean because of PCI efforts to disrupt the Marshall Plan, leading U.S. officials to fear a Communist insurrection in that country. Italy drew added attention; in 1948 it was to hold an election to select the first government under its postwar constitution. During the period prior to the election, PCI militancy increased, partly because of Italy's continuing economic difficulties and, significantly, due to international Communist criticism that the PCI received at the first meeting of the Cominform for its support of the Italian political system. The PCI's increased activity led to concern that the "Soviets were pushing ahead in Italy rather than elsewhere since Greece and Turkey were now under 'direct US protection. . . .'" American officials believed that, if the Soviet Union succeeded in establishing a Communist Italy, it would have gained a strategic position because Italy would serve to flank Greece and Turkey, and from Italy Communist influence could spread north to Germany and Austria and west into France and Spain.[6]

The Italian peninsula was deemed strategic not only for military reasons but also because from Italy a PCI government could spread communism into Europe and the Mediterranean. After Greece and Turkey were secured through the Truman Doctrine, there was concern that a PCI victory in the 1948 election or a PCI coup could cause the spread of communism back to Greece and the Middle East as well as into Western Europe. The Policy Planning Staff argued that a PCI regime in Italy would nullify the achievements of the United States in Greece and Turkey. The Central Intelligence Agency (CIA), in discussing the strategic importance of Italy and the necessity of keeping it from coming under Communist control, thought that a Communist Italy would have adverse effects in a wide area: Western Europe, the Mediterranean, and the Middle East. The CIA also stated that the availability to the Soviets of bases in Italy, Sicily, and Sardinia "would pose a direct threat to the security of US communication through the Mediterranean to Greece, Turkey and the Near East oil fields."[7]

U.S. POLICY TOWARD ITALY, 1948

Fear that the PCI could gain power in Italy, either through an electoral victory or a coup, led the United States to develop a formal policy toward that nation and to intervene in Italian electoral politics to support the campaign

[6]Alberto Tarchiani, *FRUS, 1947,* 3:970; U.S. Ambassador James Dunn to Secretary of State George C. Marshall, 7 May 1947, ibid., pp. 904–05. At the first meeting of the Cominform in Warsaw in September 1947, the PCI was criticized for its failure, symbolized by its expulsion from the Italian government. It was pressured to subordinate all party activities to the struggle against the Marshall Plan. See Donald Blackmer, *Unity in Diversity: Italian Communism and the Communist World* (Cambridge: M.I.T. Press, 1968), p. 15; and Adam Ulam, *Expansion and Coexistence: Soviet Foreign Policy, 1917–1974* (New York: Praeger, 1974), p. 449.

[7]Policy Planning Staff, 24 September 1947, *FRUS, 1947*, 3:977–81, CIA, "Italy," 10 April 1948, Carrollton Press Declassified Documents Reference System, 1978 quarterly, p. 225A.

of the Christian Democratic Party.[8] President Harry S. Truman, in December 1947, emphasized the U.S. commitment when he announced the withdrawal of American troops from Italy as required by the 1946 treaty of peace. He expressed the continued interest of the United States in an independent Italy and stated that, if Italian independence were threatened, directly or indirectly, "the United States . . . will be obliged to consider what measures would be appropriate for the maintenance of peace and security."[9] Although not expressed in so many words, Truman's statement implied that his administration would intervene with military force to prevent a Communist takeover in Italy.

In a series of policy papers known as NSC 1/1, 1/2, and 1/3, drafted in 1947 and 1948, the National Security Council (NSC) formulated U.S. policy toward Italy, pointing out that it was a significant element of national security. The NSC viewed persistent bombastic Communist propaganda attacks on the Italian government as aimed ultimately at the creation of a Communist dictatorship subservient to Moscow. In NSC 1/2, completed in February 1948, the council considered what the U.S. reaction should be if part of Italy were seized by the PCI. The council defined the basic objective of U.S. policy toward Italy as to preserve that nation "as an independent, democratic state, friendly to the United States, and capable of effective participation in the resistance to communist expansion." The NSC believed that Italy's position in the Mediterranean, since it dominated the oil supply lines to the Near East and flanked the Balkans, was essential to U.S. security because from Italy it would be possible to control Mediterranean traffic and to launch a substantial air force against any point in the Balkans. For those reasons, the NSC recommended that "the United States should make full use of its political, economic and, if necessary, military power in such manner as might be found most effective to assist in preventing Italy from falling under the domination of the USSR either through armed attack or through Soviet dominated communist movements within Italy. . . ."[10] This latter concern—the fear that Italy would fall under Communist control by a Soviet-dominated movement within Italy—eventually brought Italy into the North Atlantic Alliance.

TOWARD THE NORTH ATLANTIC ALLIANCE

The North Atlantic Treaty has its roots in the European-centered Brussels Pact, and so any consideration of the decision to include Italy must review the relationship between Italy and the Brussels powers. The failure of the

[8]See James E. Miller, "Taking Off the Gloves: The United States and the Italian Elections of 1948," *Diplomatic History* 7, no. 1 (Winter 1983): 35–55.

[9]*Public Papers of the Presidents of the United States: Harry S. Truman, 1947–1949* (Washington: Office of the *Federal Register*, National Archives and Records Service, 1963–64), p. 510.

[10]NSC 1/2, 10 February 1948, *FRUS, 1948*, 3:766–67.

London Foreign Ministers' Conference in December 1947 had led British Foreign Minister Ernest Bevin to propose the Brussels Pact. In January 1948, Lord Inverchapel, the British ambassador in Washington, explained the proposal to Secretary of State George C. Marshall, making it apparent that what Bevin had in mind was the eventual inclusion of Italy. Bevin thought that consultations should be held among Britain, France, the Benelux countries, and Italy to see what collective steps they could take to prevent extension of "the area of dictatorship."[11]

The United States was interested in the possible role of Italy in the negotiations that culminated in the Brussels Pact. Marshall contacted the U.S. embassy in Belgium in March 1948 to inquire about the views of the participating governments on the possibility of Italian inclusion and on how they contemplated meeting the problem of fifth-column aggression in Western Europe. Hugh Millard, the U.S. chargé in Belgium, reported Belgian Foreign Minister Paul-Henri Spaak's reaction was that consideration of Italy's participation would have to wait until after the 18 April 1948 election. However, Millard perceived no objection from Spaak to the eventual incorporation of Italy.[12]

The NSC urged the inclusion of Italy in the Brussels Pact negotiations as part of an effort to prevent the PCI from gaining power. In NSC 1/3, a discussion of U.S. policy toward the Italian election, the council suggested various actions to assist the Christian Democrats, among them the inclusion of Italy in discussions of regional security arrangements.[13]

Italian support for the Brussels Pact concept was initially enthusiastic but soon became hesitant. In February 1948 the Italian ambassador to the United States, Alberto Tarchiani, stated that his government welcomed Bevin's proposal even though it placed Italy in a secondary position behind France and the Benelux countries. This left the Italians to wonder whether they were truly to be included in any Western alliance and, if so, when.[14]

The Brussels Treaty was signed on 17 March 1948. It included Britain, France, and the Benelux states but excluded Italy, ostensibly because of Italy's own reluctance to join before the election. Just prior to the signing, Marshall asked U.S. Ambassador James Dunn in Rome to assess whether prompt Italian membership in the pact would be helpful to the Christian Democratic position

---

[11]Inverchapel to Marshall, 13 January 1948, ibid., p. 4; Lewis Douglas (U.S. ambassador to the United Kingdom) to Marshall, 26 February 1948, ibid., p. 32.

[12]Marshall to Belgium embassy, 3 March 1948, ibid., p. 35; Millard to Marshall, 4 March 1948, ibid., p 37. Spaak also noted that, among the participating governments, only in France was there a concern about fifth-column aggression.

[13]NSC 1/3, 8 March 1948, *FRUS, 1948,* 3:775–77. The concern that the PCI would win the election was reflected in George Kennan's recommendation which included outlawing the PCI, thereby provoking a civil war and giving the United States grounds for a reoccupation of sections of Italy. See ibid., p. 849.

[14]Conversation with Tarchiani (Norman Armour and W. C. Dowling), 18 February 1948, RG 59, 865.00/2-1848, box 6677, Decimal Files, Department of State, National Archives, Washington (hereafter cited as DSNA).

in the election. After meeting with Prime Minister Alcide De Gasperi, Dunn reported that, considering the political situation and the fact that Italy did not yet have a qualified parliament, De Gasperi believed that it would not be proper for his government to make any move toward association at that time. The prime minister did say that he thought the idea of alignment an excellent one, "particularly if it were to receive the support of the US and that he would hope eventually that Italy might become a party to it."[15]

Although the Italian government did not become a charter member of the Brussels group and wanted to wait until after the election even to consider membership, the U.S. State Department remained interested in getting Italy into the pact before the April election. The department's Policy Planning Staff urged the pact members to invite Italy to join, if De Gasperi believed such an invitation would help him in the election. After the Christian Democratic electoral victory, the official position of the United States remained that "Italy's inclusion would be natural and desirable," but it should be a decision of the treaty parties themselves.[16]

In the background of all the efforts to include Italy in the Brussels Pact, and sometimes overlapping with them, were the negotiations that led to the North Atlantic Treaty. At first, Italy's potential membership in the Brussels Pact came to be seen as a means to include it in the North Atlantic Alliance. However, as the treaty negotiations neared completion in 1949, the importance of Italy's membership in the pact diminished; discussions of its role in the North Atlantic organization intensified.

American consultations with Western European nations about a defensive association began shortly after the Brussels agreement was signed. In March 1948 the United States, Britain, and Canada began security talks in Washington. At their third meeting they agreed that the "objective is a Security Pact for the North Atlantic area, plus an extension of the Brussels agreement." The recommendation stated that "Italy would be invited to join at once if De Gasperi, in terms of Italian domestic conditions, desired such an invitation." As these conversations reached their conclusion with the sixth meeting in April, the three nations recommended that the Brussels Treaty be extended to include Norway, Sweden, Denmark, Iceland, and Italy once the election was over.[17]

Despite early indications that Italy would become a member of the new alliance, issues arose that led to a prolonged discussion by the powers. This caused the decision on the incorporation of Italy to be delayed until one month prior to the signing of the treaty. Although many issues were involved in the

---

[15]Marshall to Dunn, 11 March 1948, *FRUS, 1948,* 3:45–46; Dunn to Marshall, 16 March 1948, ibid., pp. 53–54.

[16]PPS 27, 23 March 1948, ibid., p. 62; Marshall to U.S. embassy in Paris, 6 May 1948, ibid., p. 114.

[17]Third meeting of U.S.-United Kingdom-Canada security conversations, ibid., p. 66; sixth meeting of U.S.-United Kingdom-Canada security conversations, ibid., p. 73.

talks on Italian inclusion, the debate came to center on four concerns that became increasingly interrelated: the arms limitations imposed on Italy by the treaty of peace; the fear that Italian membership would reduce the amount of U.S. arms supplied to the other powers; the initial hesitance of the Italians themselves and then their eventual determination to be included; and, most important of all, the security and geographic position of Italy.

NEGOTIATIONS

Because the Italian government had decided not to join the Brussels group before the election, the Western powers were skeptical about Italy's desire to be included in any Western defense arrangements. However, after the Christian Democrats' electoral victory, Italy began to express interest in participation in a Western alliance. In late April and early May 1948, Tarchiani held a series of discussions with U.S. officials. He pointed out that, in addition to economic and social problems, the new Italian government had to consider the issue of Italian participation in the military phase of Western European cooperation. The ambassador expressed concern about the peace treaty limitations imposed on Italy's military forces which he feared might cause Italy's membership in the pact to weaken the alliance.[18] Tarchiani asserted that Italy did not need a large military but simply one capable of defending its frontiers and making an appropriate contribution to Western European defense. He stated that his government realized that "Italy could not stand alone but must if possible enter into a scheme of Western European defense.[19]

The Christian Democrats' victory removed the problem of Communist participation in the government, but other problems emerged. Donald Maclean, first secretary of the British embassy in Washington, indicated to Theodore Achilles his belief that De Gasperi favored a relaxation of the military clauses of the peace treaty before Italy would adhere to the Brussels Pact. Achilles agreed with Maclean that Italy would present a liability to the alliance, and that it could not demand concessions in exchange for membership. However,

[18]Under the treaty of peace, Italy was forced to give up a considerable portion of its fleet, renounce possession of certain types of weapons, and to limit its army to 250,000 men and its navy and air force to 25,000 men each. Italy was stripped of all its colonies, which were turned over to the United Nations for disposition, and the city of Trieste was made into a Free Territory. See U.S., Department of State, *Treaties and Other International Acts Series 1648: Treaty of Peace with Italy*, Department of State Publication 2960 (Washington, DC: Government Printing Office, 1947); and conversation with Tarchiani, 6 May 1948, *FRUS, 1948*, 3:797.

[19]Conversation with Tarchiani, 6 May 1948, *FRUS, 1948*, 3:797.

Achilles pointed out that there was room within the treaty limitations to strengthen greatly the Italian armed forces.[20]

In July the Rome government appealed to the United States to encourage the Brussels nations to include Italy. Tarchiani told Undersecretary of State Robert Lovett that it was not worthwhile to build up Italy economically without taking any steps to prevent a Soviet takeover. The ambassador argued that the Western Union powers looked to the United States for guidance on military matters and would accept an American suggestion that Italy join them. He said that "the United States is the key to the defense of Europe and the association in question is not so much one of a European as it is of an 'Atlantic' community. Consequently, the United States has every right to express the American viewpoint on matters which affect American security."[21] Despite Italian expressions of interest in joining the Brussels Pact, doubts persisted about their sincerity. These doubts were due to a number of factors. During the election campaign a great deal of emphasis was placed on the need for peace, perhaps indicating that many Italians might take a neutral position in the Cold War. Others who may have favored adherence argued that Italy should demand peace treaty revision as a price for membership. Still others feared that Italy would become entangled in an alliance in which its military weakness would severely reduce its influence.[22]

Because of these doubts the Italian government found it necessary to combat the impression that Italy did not desire membership or would demand concessions, such as a relaxation of the military clauses of the peace treaty, as its price for membership. De Gasperi made it clear to Dunn that his country was not neutral; it was in the Western group, as the election indicated. Italy, the prime minister said, was aligned with the Western states "from [the] point of ideology and common objectives of independence and individual liberty." The problem for his government was to counteract the strong PCI effort to discourage Italy from engaging its future with the West.[23]

Despite De Gasperi's efforts, doubts still remained. Spaak indicated that, in a November meeting with De Gasperi, the prime minister had been "cautious and reserved" about Italian participation because national sentiment had not yet crystallized in favor of NATO. However, contrary to Spaak's apprehension that De Gasperi would raise the issue of treaty revision, the prime minister told him that "it was premature to worry about Italy's military

---

[20]Conversation with Maclean, 7 May 1948, ibid., pp. 115–16. The problem was that the only way to alter the provisions of the treaty was by mutual consent of the signatory powers or, after Italy entered the United Nations, through Security Council action. The Soviet Union refused to agree to alterations and vetoed Italian membership in the United Nations on several occasions. Eventually, in 1952, Italy and the Western powers abrogated most of the limitations despite Soviet objections. See Dean Acheson to Tarchiani, 21 December 1951, U.S., Department of State, *Bulletin* 31 (December 1951): 1050.

[21]Conversation with Tarchiani, 26 July 1948, *FRUS, 1948,* 3:200–01.

[22]John Hickerson to acting secretary of state, 7 October 1948, ibid., pp. 260–61.

[23]Dunn to Marshall, 15 September 1948, ibid., pp. 252-53.

strength since only five divisions of Italy were currently in existence out of [a] possible total of twenty."[24]

When the U.S.-United Kingdom-Canada security conversations broadened into the Washington Exploratory Talks on Security by including the Brussels powers, Italy became a major issue. At this stage in the negotiations Italian inclusion became a significant controversy because its geographic position placed it outside the North Atlantic area. A September memorandum of the Washington talks discussed the territorial scope of the pact and assessed the problem in military terms:

> A North Atlantic security system composed exclusively of the United States, Canada and the present parties to the Brussels Treaty would not be fully effective. On the other hand even the combined military resources of these nations would be inadequate to warrant their assuming hard and fast commitments for the security of a large number of geographically scattered countries. A line must be drawn somewhere. The problem is to devise an arrangement which would best meet the security needs of the nations here represented without overextending their military capabilities.[25]

The memorandum also outlined a number of factors that would be important in the consideration of where that line would be drawn: the effect that a threat to the political or territorial integrity of a state would have on the nations involved in the discussions, the necessity of strengthening the Western orientation of some countries, and the importance to the West of specific nations under consideration. It was noted that the case of Italy presented certain difficulties because it was not a North Atlantic country and was subject to military limitations imposed by the treaty of peace. However, Italy's territory was of strategic importance to the nations involved, and its Western orientation had to be maintained and strengthened. "A satisfactory solution of the problem" had to be found.[26]

To the participants in the North Atlantic Treaty negotiations, Italy presented a perplexing paradox. Military opinion in both Europe and the United States held that Italy could make little contribution to security and, in fact, might be a drain on military strength. As John Hickerson, director of the State Department's Office of European Affairs, noted, the reason motivating Benelux and French opposition to Italian participation was the "desire to avoid having their share of the U.S. arms pie reduced by cutting another slice." Yet as Dean Acheson wrote, "from a political point of view an unattached

---

[24]Alexander Kirk (U.S. ambassador to Belgium) to Marshall, 29 November 1948, RG 59, 865.20/11-2948, box 6688, DSNA.

[25]Memorandum by participants in Washington security talks, 9 September 1948, *FRUS, 1948,* 3:240–41.

[26]Ibid., p. 241.

Italy was a source of danger." He feared that if Italy were excluded it would suffer an isolation complex, encouraging its seduction by the Soviet Union.[27]

Although no official American position existed, it appeared that the United States was not interested in entering a pact with the Brussels parties alone and attached great importance to a "provision for [the] original membership of Italy." Lovett noted that, although Italian inclusion in the Brussels Pact was a matter for decision by the involved parties, the United States did have strong views "on initial inclusion [of] Italy in any Atlantic Pact." Yet a month later, at the tenth meeting of the Washington Exploratory Talks in December, he stated that his country "had no firm position" at that time. The undersecretary of state also noted that the U.S. attitude toward Italian inclusion in the North Atlantic Treaty negotiations probably would depend on the position of the Brussels Pact powers toward Italian membership in that alliance. He did not believe that Italy should be left without some sort of protection, but neither did he then attempt to suggest any solution to the problem.[28]

The European powers, with the exception of France, opposed Italian inclusion. The Belgian representatives tended to view the security of Italy as an essentially Mediterranean problem, and they thought that Italy should eventually be protected by some form of Mediterranean security arrangement. The Canadians believed that commitments should be limited to nations bordering on the North Atlantic. The other states participating in the talks expressed concern about Italian security but did not favor its charter membership in the new pact.[29]

Great Britain led the opposition to Italian membership. The British view, which was generally shared by the others, was based on the belief that Italy would be not only more of a military liability than an asset but also a drain on the limited military resources available. In addition, they thought it illogical to invite a non-Atlantic power and feared that it would lead to questions relating to the peace treaty. The British were aware of the strategic importance of Italy. General Sir Leslie Hollis, then deputy secretary (military) to the cabinet, admitted that Italy, Sicily, and Sardinia were of great importance to the control of Mediterranean communications and should, at the least, be kept in neutral hands. Therefore, he argued that the Italians should be armed and be given assistance in times of war. However, Hollis noted:

> The British Chiefs of Staff are of the firm opinion that it would be fundamentally unsound at this stage to add to our existing commitments. . . . Moreover, they would be most reluctant to see armaments

[27]Hickerson to William J. McWilliams of the NSC, 24 November 1948, RG 59, 840.20/11-2448, box C–510, DSNA; Dean Acheson, *Present at the Creation: My Years in the State Department* (New York: W. W. Norton, 1955), p. 279.

[28]Lovett to U.S. embassy in Belgium, 22 November 1948, *FRUS, 1948*, 3:282–83; minutes of tenth meeting of Washington Exploratory Talks, 22 December 1948, ibid., p. 330.

[29]Alexander Kirk to acting secretary of state, 7 January 1949, *FRUS, 1949*, 4:14; minutes of tenth meeting of Washington Exploratory Talks, 22 December 1948, *FRUS, 1948*, 3:327.

> manufactured in the United States, that would otherwise go to countries of the Western Union, going to Italy instead. In short, they strongly support the Foreign Secretary's view that 'at this stage Italy would be a military liability.'[30]

Other European concerns were that if Italy were included it would be difficult to keep out Greece and Turkey, and that Italian membership would weaken the alliance "by making it less homogeneous . . . and it would be more difficult to use the alliance as a chrysalis for a North Atlantic community."[31]

The Italians eventually found a firm supporter in France. Initially, the French had opposed Italian membership and desired the Brussels powers to limit themselves to dealing with the security problems of Western Europe by reinforcing the pact through association with the United States and Canada. Instead of Italian participation, they favored the protection of Italy by the creation of a separate defense system that would include the Mediterranean and the Persian Gulf.[32] The French government, however, came to view the security of Italy as essential for its own security because the two countries shared a common border. The French were willing to concede that Italy was not a North Atlantic nation, but they argued that it belonged to the North Atlantic "security area." In particular, the French military supported Italian incorporation because they were convinced that the defense of Italy, particularly the Po Valley, was essential to the defense of France. To them, effective defense arrangements for Western Europe without Italy were unthinkable.[33]

Henri Bonnet, the French representative to the Washington talks, continually countered opposition arguments. He stated that "the argument based on the limitation of Italian armaments in the peace treaty was not serious, as precautions could be taken to ensure that Italy conformed to the treaty." The French also argued that bringing Italy into the pact need not lead to membership for Greece and Turkey. The crucial point, said Bonnet, was that between Italy and the Brussels Pact states there existed a territorial continuity, while no such continuity existed between Greece and Turkey and Western Europe.[34]

Aside from the obvious French concern with their security, the other nations involved in the discussions believed that there was another motive

---

[30]Foreign Office to Rome, 13 January 1949, F.O. 371/79220, 8595, 7260/1074/72, Public Record Office, London (hereafter cited as PRO); General Sir Leslie Hollis to Gladwyn Jebb, 20 January 1949, F. O. 371/79222, 8619, GR 2681, Z107, C.O.S. 123/20/1/9, ibid.

[31]Escott Reid, *Time of Fear and Hope: The Making of the North Atlantic Treaty, 1947–1949* (Toronto: McClelland and Steward, 1977), p. 200.

[32]Henri Bonnet, fourth meeting of Washington Exploratory Talks, 8 July 1948, *FRUS, 1948*, 3:165; Jefferson Caffery (U.S. ambassador to France) to Marshall, 24 June 1948, ibid., p. 143; Reid, *Time of Fear*, p. 201.

[33]Memorandum by William Sale, 2 November 1948, RG 59, 865.20/11-248, box 6688, DSNA.

[34]Tenth meeting of Washington Exploratory Talks, 22 December 1948, *FRUS, 1948,* 3:329; eleventh meeting of Washington Exploratory Talks, 14 January 1948, *FRUS, 1949,* 4:29.

for French support of Italian inclusion: France's desire to protect its interests in North Africa. The British believed that the French thought "that the inclusion of Italy would make the Anglo-Saxon powers more favorable towards the idea of French North Africa being included in the Geographic scope of the treaty."[35] Indeed, Escott Reid has asserted that the French began to push for Italian membership because they saw that, once a purely Mediterranean nation such as Italy was included, it would be easier to extend the North Atlantic Treaty to North Africa. If they could gain protection for Algeria, many believed that the French would then abandon their support for Italy.[36] Hume Wrong, the Canadian representative to the talks, thought the entire issue of North Africa depended on what was done about Italy. He wrote that if Italy were included "there would be a good case on geographical ground for including the Mediterranean coast of North Africa.[37]

In early 1949 it remained doubtful that Italy would be asked to join. France, albeit for selfish reasons, supported Italian membership, and it seemed that the United States did also, but the other nations involved in the pact negotiations remained steadfast in their opposition.

As discussions continued, the Italian government representatives themselves began to play a more active role. De Gasperi and Italian Foreign Minister Carlo Sforza were determined that Italy be included in either the Brussels Pact or the Atlantic Alliance. The major difficulty they encountered was in getting the support of the Social Democrats and their leader Giuseppe Saragat.[38] Sforza and De Gasperi thought that the continuation of a coalition government with the Social Democrats was essential to the Christian Democrats' program, and that Saragat was important because he was the government's strongest link to the labor movement. However, Saragat did not believe that it was necessary at the present time to take such a definite step as joining the Western alliance. In order to alleviate Saragat's concerns, De Gasperi met with him in early January 1949. The prime minister explained that Italy's entrance into the alliance would virtually guarantee U.S. aid in case of Soviet aggression and emphasized that the choice was either security through a U.S. guarantee or an isolated, defenseless Italy.[39] After a meeting with all political leaders, except those of the Communist and Socialist parties, the Italian government fully agreed upon the desirability of entering into defense arrangements, preferably the North Atlantic organization.

On 6 January the Italian government informed Washington of its decision. A request was formally presented on 12 January by Ambassador Tarchiani,

---

[35]Foreign Office to Rome, 13 January 1949, F.O. 371/79220, 8595, Z260/1074/72, PRO.

[36]Reid, *Time of Fear*, p. 203.

[37]Minutes of tenth meeting of Washington Exploratory Talks, 22 December 1948, *FRUS, 1948*, 3:327.

[38]The Social Democrats, officially known as the Socialist Party of Italian Workers, broke off from the Socialist Party in 1947 over the continued Socialist alliance with the PCI.

[39]Dunn to acting secretary of state, 5 January 1949, *FRUS, 1949*, 4:7–8; *New York Times*, 7 January 1949.

who emphasized that his country would prefer to join the Atlantic Alliance rather than the Brussels Pact because the Italian public looked at the former as something concrete, whereas the latter was regarded as only a good intention. The ambassador warned of the possible effects if Italy were not included. He said that the Communists "would have a field day," and that the result would be increased unrest and the destruction of any hope for economic recovery and political stability in Italy.[40]

The request put pressure on the United States, which had not yet taken an official stand on the participation of Italy. Hickerson noted that all nations "generally felt that this request would make it much more difficult not to include Italy." The British also were aware of its importance. Sir Oliver Franks realized that, although his government's opposition had not been altered, "the formal approach on the part of the Italian government did change the picture and would necessitate reconsideration of the question by his Government."[41]

In addition to their formal request, the Italians lobbied in support of their application. The Italian ambassador to Belgium visited Alexander Kirk in order to assure him that the Italians were sincere in their desire to join the pact. In addition, the Italian ambassador to Canada, Laurence Steinhardt, called on Prime Minister Lester Pearson to persuade him to withdraw Canadian objections. He told the prime minister that if Italy were denied membership the adverse reaction would be so severe that it could "endanger tenure of [the] present Government."[42]

As a result of the Italian application, the emerging NATO powers increasingly came to lay the burden of the decision on the United States. Spaak thought that the request was ambiguous, but if the Italians took a categorical position, he would support their initial adherence. He added that "in the final analysis [the] question lay between Italy and the US and that if [the United States feels] . . . strongly that Italian inclusion is essential he will certainly not oppose this."[43] On 28 January 1949, at the Consultative Council meeting of the Brussels powers in London, it was agreed that Italy could join, if the United States insisted.[44] Thus the Europeans deferred to American wishes.

---

[40]*FRUS, 1949*, 4:23–24.

[41]Hickerson to Lovett, 14 January 1949, RG 59, 840.20/1-1449, box C–511, DSNA; eleventh meeting of Washington Exploratory Talks, 14 January 1949, *FRUS, 1949*, 4:29.

[42]Kirk to Acheson, 26 January 1949, RG 59, 840.20/1-2649, box C–511, DSNA; Steinhardt to Acheson, 27 January 1949, RG 59, 840.20/1-2749, box C–511, DSNA.

[43]Kirk to Acheson, 22 January 1949, RG 59, 840.20/1-2249, box C–511, DSNA.

[44]Hickerson to Acheson, 11 February 1949, RG 59, 840.20/2-1149, box C–511, DSNA. Acheson noted that at this meeting there was some confusion about the decision. It was reported to the Department of State that, if the United States insisted, the other countries would not object. Acheson said, however, that he believed they had agreed that they wanted Italy in NATO. See U.S., Congress, Senate, Committee on Foreign Relations, *North Atlantic Treaty: Hearings Before the Committee on Foreign Relations,* 81st Cong., 1st sess., 1949, 1:164.

THE UNITED STATES DECIDES

Within the American foreign policy establishment there were great differences of opinion on the Italian question. Several alternatives were under consideration: 1) inclusion of Italy in the alliance alone; 2) inclusion in the Brussels Pact, with simultaneous inclusion in the Atlantic Alliance; and 3) exclusion from both, but an announcement by the United States, Britain, and France that an attack on Italy would be an occasion for consultation under the North Atlantic Treaty. However, one State Department official, William Sale, believed that the third option was no option at all. He wrote: "I think it should be made clear to the Secretary that the alternative to Italian participation in the pact would be a neutral Italy. . . ."[45]

Within the State Department the conflict over Italian membership in the alliance centered on the issues of geography and security. Two key individuals were involved in the debate: Hickerson, who supported inclusion, and George Kennan, who opposed Italian membership. Kennan not only opposed Italian inclusion but, along with State Department counselor Charles Bohlen, also opposed the entire concept of the North Atlantic Treaty.

Kennan's objections to the treaty were based on his perception of European events. He thought that it was necessary to consider possible changes in Europe which, in the long run, could lead to a general unification of the entire continent. He was concerned that "the door should be left open for everyone in Europe to come in . . . so that there could be a real unification of Europe." He also believed that discussions should not focus only on the question of a military guarantee against a military danger that he hoped would be eventually mitigated.[46] In general, Kennan feared that the North Atlantic Treaty would cause a "preoccupation with military affairs, to the detriment of economic recovery and of the necessity for seeking a peaceful solution to Europe's difficulties."[47]

Once he had accepted the fact that the alliance would be formed, Kennan then was concerned with its territorial scope, which led him to oppose Italian inclusion. He believed that the United States would be on solid ground only if the treaty was called a North Atlantic security pact and if its membership were strictly limited to the North Atlantic area. He thought that nations like Italy should be protected but argued that this could be done by announcing, at the conclusion of negotiations, that if any other European Recovery Program state was threatened it would be regarded as an occasion for consultation by the NATO powers. This, according to Kennan, would exclude Greece and

---

45Lovett to Dunn, 5 January 1949, *FRUS, 1949*, 4:8; memorandum by WBS [Sale], n.d., RG 59, Records of the Office of Western European Affairs Relating to Italy, box 1, DSNA.

46Kennan, minutes of fifth meeting of Washington Exploratory Talks, 9 July 1948, *FRUS, 1948*, 3:177.

47George Kennan, *Memoirs, 1925–1950* (Boston: Little, Brown and Company, 1967), pp. 402–10.

Turkey, "and probably Italy as well." His rationale for a limited geographic scope for the treaty was based on the necessity of avoiding the appearance of aggressive encirclement of the Soviet Union. Kennan thought that only a sound geographical standard for membership could clearly demonstrate the defensive nature of the alliance.[48] In his memoirs he stated that he did not advocate ignoring Italy and noted that continued U.S. military presence in Italy was necessary until the Trieste and Austrian peace treaty problems had been solved.[49]

Kennan's point of view was reflected in a Policy Planning Staff paper entitled "Considerations Affecting the Conclusion of a North Atlantic Security Pact" (PPS 43). This paper asserted that the North Atlantic Treaty was "not the main answer to the present Soviet effort to dominate the European continent." The staff also supported Kennan's view on the territorial scope of the treaty:

> The Policy Planning Staff is of the opinion that the scope of a pact of this sort should be restricted to the North Atlantic area itself, and that attempts to go further afield and to include countries beyond that area might have undesirable consequences.
>
> The Staff considers that a North Atlantic security pact might properly embrace any country whose homeland or insular territories are washed by the waters of the North Atlantic, or which form part of a close union of states which meet this description.[50]

On the other hand, Italy had a strong backer in the person of Hickerson, who spelled out his rationale for support in two memorandums written in response to PPS 43.[51] Hickerson noted that he agreed with the thesis that the territorial scope should be limited to natural members, but he believed that Italy was one of those natural members. He stated that the NSC decided that a Soviet attack on Italy would involve the United States in a war, and emphasized that Italy's exclusion would be interpreted by the Italians as an indication that their country was not an integral part of the Western Union concept, possibly

---

[48]Kennan to Lovett, 31 August 1948, *FRUS, 1948*, 3:225; Kennan, *Memoirs*, p. 411.

[49]Kennan wrote that, once the question of a military alliance had been raised, it became difficult to exclude Italy without "having misleading implications that could affect in unfortunate ways the delicate balance of the domestic-political struggle within that country." He believed that the United States should have confined its actions to a guarantee of the Brussels Pact powers and a military assistance program designed to boost morale, thereby avoiding the entire Italian problem. See Kennan, *Memoirs*, pp. 412–13.

[50]PPS 43, 23 November 1948, *FRUS, 1948*, 3:284–85.

[51]Reid, in *Time of Fear*, argues that Hickerson played a significant role in the inclusion of Italy. Hickerson recently said that Reid "completely exaggerated" his role. Hickerson stated that he pushed the admission of Italy simply because he "felt it would be a tragedy to leave out a country who has contributed so much to our culture and well-being . . . and feared that Italy would be taken over by communist-backed Italians if Italy were left out in the cold." Personal letter to the author from Hickerson, 11 April 1983.

weakening resistance to communism.[52] In a second response to PPS 43, Hickerson wrote directly to Kennan, agreeing with his basic thoughts about limitations on the territorial scope of the pact but differed on the definition of the area: "The North Atlantic 'area' is construed . . . as including Italy. . . . Italy must be included. I would consider that Italy, . . . Sweden and Germany were natural members of the Atlantic community."[53]

In his efforts to gain Italian inclusion, Hickerson also put pressure on Secretary of State Acheson. Hickerson recognized that the United States had sought to avoid being placed in the position of including Italy at its insistence and believed that the Europeans must agree on a solution. He suggested that Italy join the alliance through the Council of Europe. Hickerson favored this because he believed that "inclusion of Italy in the declaration contemplated for Greece, Turkey and Iran would not be adequate."[54]

Hickerson also had the support of the Joint Chiefs of Staff. The military establishment, with certain reservations, favored Italian inclusion in the pact. In commenting on the territorial scope of the treaty, Admiral Louis Denfeld feared that extension of the treaty to areas of the Mediterranean would "unnecessarily and dangerously" broaden U.S. responsibilities. The admiral insisted that the primary reason for Italian inclusion was political. In his memorandum to the secretary of defense, he noted that there was no military necessity for Italy to be in NATO. However, he stated that from the military point of view it was desirable for Italy to be in the Brussels Pact because it was essential that the defense of Western Europe be integrated and more compactly drawn. By including Italy, Denfeld argued that the greatest economy of forces could be obtained. He stated that "it would, in fact, be unrealistic militarily for Italy to be a member of neither." He argued that this was the case because, in terms of land warfare in Western Europe, Italy was strategically important, and, in terms of sea warfare, there was no question concerning its strategic potentiality for the control of the Mediterranean. The admiral concluded that it would be harmful for Italy not to be a member of either the Brussels Pact or the Atlantic Alliance.[55]

Along with the strategic importance of Italy, expressed by both the Joint Chiefs of Staff and the NSC, the military also had political concerns about Italy. General Omar Bradley told British military leaders that if Italy were excluded they would lose heart and would "probably be ripe to accede to

---

[52]Hickerson to McWilliams, 24 November 1948, RG 59, 840.20/11-2448, box C–510, DSNA.

[53]Hickerson to Kennan, 26 November 1948, RG 59, 840.20/11-2648, box C–510, DSNA. The idea of Italy being a natural member of the North Atlantic area was also included in the December 1948 draft treaty annex on Italy. In the annex it was noted that, if one aim of the pact was to tighten cultural and political ties, then Italy, by culture, civilization, mercantile, and maritime traditions, would be an appropriate member. See draft treaty, *FRUS, 1948*, 3:340–41.

[54]Hickerson to Acheson, 24 February 1949, RG 59, 840.20/2-2449, box C–511, DSNA.

[55]Memorandum by Joint Chiefs of Staff to James Forrestal, 5 January 1949, *FRUS, 1949*, 4:12–13.

Soviet demands on outbreak of war, and may even meanwhile go communist."[56] Overall, the joint chiefs believed that Italy was in an exposed position, that it was subject to continuing Communist pressure, and that it would be a source of weakness if left out of the alliance.[57]

Hickerson's pressure on Acheson, backed by the opinion of the military and the NSC, finally led the secretary to meet with Truman in order to determine an official position. On 28 February 1949 the president and secretary of state discussed the problem of Italy. At that meeting, Truman had no final opinion on the issue and asked Acheson to talk to certain senators. At the time, Truman was inclined to think that it might be wiser not to have Italy included as an original signatory to the treaty, and possibly not in NATO at all. He would have liked to consider the possibility of some sort of separate Mediterranean arrangement.[58]

Before returning to the White House for a final decision, Acheson met with Senators Tom Connally, Arthur Vandenberg, and Walter George. All three opposed Italian membership, fearing that the United States might overcommit and overextend itself militarily and financially. However, when reporting to Truman, Acheson noted that the senators believed that responsibility for such a decision rested with the executive, and they would not oppose any decision once it had been made.[59]

Finally, on 2 March 1949, Acheson met again with the president. He presented Truman with a memorandum that outlined the arguments for and against the incorporation of Italy. The paper, which supported Italian membership, emphasized the danger of exclusion. It noted that the United States had devoted much effort to strengthen Italy's Western orientation and to weaken the Communist threat in that nation. It also commented on the fact that the Italian government had made great progress in preparing the public for closer cooperation with the West and had asked to be included in the alliance. The memorandum warned that "Ambassador Dunn feels very strongly that a rebuff to Italy in this connection . . . would substantially nullify all that has been gained in Italy during the past year and lead the Italian Government to adopt a position of neutrality. . . . A rebuff would increase the Communist influence in Italy and discredit the present government and its pro-western policies." Acheson informed Truman that the French and Canadians supported inclusion, while the Netherlands, Belgium, and Luxembourg

---

[56]Hollis to Jebb, 20 January 1949, F.O. 371/79222, 8619, GR 2681, Z107, C.O.S. 123/20/1/9, PRO.

[57]Acheson, 8 March 1949, U.S., Congress, Senate, Committee on Foreign Relations, *The Vandenberg Resolution and the North Atlantic Treaty: Hearings Held in Executive Session Before the Committee on Foreign Relations*, 80th Cong., 2d sess., 1973, p. 164.

[58]Memorandum of conversation with Truman (Acheson), 28 February 1949, *FRUS, 1949*, 4:125.

[59]Sarah Jane Gilbert, "From Armistice to Alliance: Goals and Methods in Italian Foreign Policy" (Ph.D. diss., Johns Hopkins University, 1970), p. 495; Reid, *Time of Fear*, p. 208; memorandum by Acheson, 2 March 1949, *FRUS, 1949*, 4:145.

had no objections. The secretary also told the president that the United States was in a position of either accepting the European position or rejecting it, warning that such a rejection would result in serious difficulties with France, cause delay in the conclusion of the treaty, and generate much publicity about a split among the Western powers.[60]

The fear of discord, in fact, was quite real. The French ambassador to Britain had informed Bevin that there would be little hope of getting the pact through the French parliament "if Norway was a member of it and Italy was not." If Italy were excluded, the French argued, they were being asked to defend places such as Iceland and Alaska with no guarantee that the United States would defend France in case of a Soviet invasion into the Po Valley, which the French considered their "doorstep." For that reason, the French were unwilling to contemplate the signature of the treaty without Italian membership.[61] Thus, it can be argued that the United States decided to include Italy in the alliance, but the French position played a major role in that decision.

Acheson recommended that he be authorized to agree to Italian inclusion, and Truman accepted that recommendation, although he still preferred a treaty without Italy.[62] At the fifteenth and sixteenth meetings of the Washington talks, Acheson informed the other nations that the United States could now agree to an invitation to Italy to participate in the pact.[63] On 8 March, Hickerson met with Tarchiani and extended an invitation for Italy to participate in the final discussions as to become a signatory member of the North Atlantic Treaty.

CONCLUSION

After months of negotiations Italy became a charter member of the alliance. With a war-devastated economy, an armed forces limited by treaty, and a shoreline washed by the Mediterranean, it was included in the North Atlantic Treaty. In large part, Italy's membership was the result of the fear that a Soviet-backed fifth column could cause the loss of a strategically important friendly territory. Although the NSC, the military, and some State Department officials stressed the strategic importance of Italy, Hickerson did not. To Hickerson, the major reason for the incorporation of Italy was the belief that if Italy were left out that nation would fall to the Communists. In

---

[60]Memorandum by Acheson, 2 March 1949, *FRUS, 1949*, 4:145, 141–42.

[61]Ernest Bevin to Harvey, 2 March 1949, F.O. 371/79231, 8710, Z1991/1074/72G, PRO.

[62]Memorandum by Acheson, 2 March 1949, *FRUS, 1949*, 4:142.

[63]Fifteenth and sixteenth meetings of Washington Exploratory Talks, 4 and 7 March 1949, ibid., pp. 151–52, 167. It was noted at the fifteenth meeting that "the inclusion of Algeria in the Pact . . . had strengthened Italy's case."

retrospect, when the treaty was signed he downplayed the strategic importance. He thought that after 1949, and especially after Greece and Turkey were admitted, the strategic importance of Italy, with its naval bases for the Mediterranean fleet, was more fully recognized. Hickerson has stated that in 1949 he believed that it was a matter of U.S. national interest: "I hated to see one of the mothers of our civilization go communist when I felt that it could be avoided. . . . We won the election [April 1948] and I hated to see that lost by having them left out of the treaty."[64]

The countries that negotiated the treaty feared the possibility of indirect aggression by the Soviet Union through the Communist parties of Western Europe. The Western governments believed that the greatest danger in 1948–49 was not a Soviet invasion of Western Europe but a continued subversion of Western European governments by Soviet-backed internal fifth columns. Given the U.S. perception of the PCI, and indeed all Communist parties as agents of the Soviet Union, any action taken by the Italian Communists would be assumed to be assisted or instigated by the Soviets. Hickerson, speaking thirty-one years after the signature of the treaty, said that he believed it was the intention of the treaty to protect nations against subversion and that, if the PCI tried a coup after the signing of the treaty, the NATO powers would be compelled to assist the Italian government. He further stated that "it is not clearly provided for but I think by implication it probably is."[65] American policy in 1948 toward a civil conflict in Italy lends support to Hickerson's view. According to NSC 1/2, if part of Italy were to fall to the Communists, the United States would stop all aid to Communist-controlled areas, strengthen the American military establishment, bolster U.S. military forces in the Mediterranean, provide a show of strength in non-Communist areas of Italy, and deploy U.S. forces to Sicily or Sardinia.[66]

Such a use of the North Atlantic Treaty, and the U.S. role in it, is revealed in NSC 67/1 of April 1950. This policy paper states that the Italian government would require assistance in suppressing an internal Communist insurrection, and that "Italy may be expected to invoke the appropriate provisions of the North Atlantic Treaty."[67] Even though there was no direct evidence in 1950 that the PCI intended an insurrection, the North Atlantic alliance was to be used to prevent the PCI from coming to power in Italy and

[64]Personal interview with Hickerson, 18 June 1980. Hickerson's belief that exclusion would lead to a Communist Italy was also indicated in a memorandum he wrote in 1948. He stated that "it would be a tragedy if after winning the first round in the March [*sic*] election we should lose the second." See Hickerson to McWilliams, 24 November 1948, RG 59, 840.20/11-1448, box C–510, DSNA.

[65]Personal interview with Hickerson, 18 June 1980. After hearings that included testimony by Acheson and General Bradley on indirect aggression and the role of the North Atlantic Treaty, the Senate Foreign Relations Committee concluded that consultations could be sought in the case of an internal disorder where there was indication that such disorder was assisted from outside the nation involved. See *Vandenberg Resolution and the North Atlantic Treaty*, p. 155.

[66]NSC 1/2, 10 February 1948, *FRUS, 1948*, 3:765–69.

[67]NSC 67/1, 21 April 1950, *FRUS, 1950*, 3:1487–89.

thereby keep that strategic nation—located between U.S. interests in Western Europe and its interests to the East—within the Western bloc.

Beginning after the Second World War, U.S. concern about the security of Italy caused American policymakers to conclude that Italy was linked to the security of both the Mediterranean and Western Europe. As the Western European governments began the negotiations that eventually led to the North Atlantic Treaty, the United States remained concerned about the position of Italy in Western defense. With Italy in NATO, there was a mechanism created that could be used to prevent the PCI from taking power. Within NATO, an intervention would no longer have to be unilateral and could be portrayed as a united effort of the Western powers in preventing the spread of communism outside its sphere of influence. With the inclusion of Italy, the treaty made the security of the Mediterranean a major area of concern to the West.

The Italian Role in NATO

RAIMONDO LURAGHI

The Mediterranean is a world in itself. Not only was Western civilization born and developed along its shores; not only did Marco Polo and Christopher Columbus depart from there; but today its basin is also "the only place on the globe where the 'four worlds'—Western democracies, Communist regimes, the prospering oil producers and the poor developing countries—live in such close neighborhood."[1]

On its shores the world passes in review. The Mediterranean, in fact, ". . . embraces about every conceivable issue that is on mankind's agenda today: oil, food, inflation, unemployment, social unrest, minority problems, the crisis of institutions, arms trade and armed conflict, potential nuclear proliferation, pollution and an emerging struggle for ocean resources."[2] It is indeed impossible to overestimate the importance of the "internal sea."

Communications

The Mediterranean has always been, is, and remains the route par excellence of trade between East and West. Hundreds of merchant ships per day—1,200 belonging to NATO countries—ply its waters. The disruption of that trade could deal a fatal blow to countries like Greece and Italy.

Population

More than 300 million people live on its shores. The region's armed forces include about 3 million men at arms.

Minerals

These range from crude oil to bauxite, phosphates, lead, copper, zinc, chromium, and sulfur.[3]

This all contributes to making the Mediterranean problem one of the main concerns for strategists and governments, but this is not enough. Since

---

[1]Curt Gasteyger, "Superpowers in the Mediterranean," *Atlantic Community Quarterly* 14 (Spring 1976): 49.

[2]Ibid.

[3]Ihsan Gürkan, *NATO, Turkey and the Southern Flank: A Mideastern Perspective* (New Brunswick, NJ: Transaction Books, 1980), p. 6.

the era of the wars between Egyptians and Hittites, most of the decisive battles in history—from Marathon to Lepanto, from Zama to El Alamein, from Abukir to Vittorio Veneto—have been fought in countries surrounding the Mediterranean. With the exception of the brief period in which the "Pax Romana" dominated the whole sea, it always has been the theater of struggles between divided and adversary civilizations. For Rome, the attempt made by Caesar, and vainly pursued by Augustus, to "cover" the back of its own Mediterranean empire by means of a firm grasp of continental Europe did not succeed. Since Rome failed in its attempt to conquer Germany, its destiny—and with it the future of Mediterranean unity—were doomed.

STRATEGIC PROBLEMS

What is the strategic lesson to be learned from such experiences? First, that the Mediterranean and Central-Western Europe either will stand or fall together. Having lost the Elbe and Rhine positions, it would be impossible for the Mediterranean world to withstand the pressure of the European continental mass, but at the same time the Mediterranean always has been, and is now more than ever, the route for the potential outflanking of Central Europe. The fact that an enemy might take possession of its shores, and that such an enemy could easily break up Mediterranean defense lines, might mean that Central Europe would become, in turn, indefensible with appalling prospects.

In this case, the first Western defense line would be the Atlantic Ocean. It would no longer be possible even to think of holding the English Channel, as was done from 1940 to 1944, given the power of today's weapons. So the Mediterranean emerges in today's NATO defensive strategy as a crucial theater. Although the Mediterranean was already of primary importance when the Atlantic Alliance was created, it is significant that the association was named precisely "Atlantic." The Mediterranean in 1949 was considered of secondary importance to the defense of the West. In today's situation, when Western security margins have dangerously narrowed, it has become a decisive sector, a point that not many politicians seem to understand.

The first problem is created by the Soviet naval presence in the Mediterranean. According to Edward Luttwak and Robert Weinland,

> On the face of it, the inflexibility of Russian naval power is a thing of the past. Much-strengthened submarine forces and modernized naval-air forces have now been joined by a forward-deployed surface fleet of missile destroyers and cruisers and helicopter carriers. On the face of it, this new Russian naval strength is such that (a) it can negate American naval tactical options and thus undermine the credibility of American naval

> suasion, and (b) it can exercise naval suasion as an instrument of Soviet foreign policy.[4]

Obviously the Soviet navy operates in the Mediterranean under a series of restrictions. For the time being, all accesses to this sea, from Gibraltar to Suez to the Turkish Straits, are controlled by NATO powers or by a friendly state. Soviet strategists have to face the problem of a series of ramparts not easily overcome. These constitute highly useful tools in the hands of the Atlantic leaders, provided they are able to use them.

TURKEY AND THE SOVIET UNION

The first rampart, whose highest importance is usually understated, is the Turkish-Anatolic stronghold. Not only does Turkey have control of the Straits, but it also interposes itself as a barricade that closes access entirely, both directly from the Black Sea and through the Transcaucasian region. It was not by chance that the eastern Roman Empire, when everything seemed to collapse, succeeded in beating back the barbaric waves, retreating into such a natural stronghold, and in resisting for another thousand years. This was also the natural fortress of the Turkish nation after the collapse of 1918. It suffices to take a glance at a historical atlas to understand how both strongholds—Byzantine and Turkish—having lost in both cases the western and eastern sections, enclosed themselves more or less within the same limits, from the Turkish Straits to Mount Ararat and from the Black Sea to the gates of Syria. For this reason, the situation in Mesopotamia and in Iran has always been important for the defense of Anatolia. For this same reason, NATO must follow carefully the evolution of events in Iran, which throughout many turmoils today appears more vulnerable than ever.

Within this zone lie countries that for centuries have been pivots of Russian and then Soviet imperialism: Syria, Iraq, Azerbaijan. With its massive presence in Syria, the Soviet Union has been able, to some degree, to outflank the Turkish rampart. In the meantime, Syria, with its harbors of Latakia and Tartus, has become one of the main ports of call for the Soviet Mediterranean squadron. By 1976 Soviet naval ships were spending a yearly average of 340 days there.[5]

The current war between Iran and Iraq has certainly complicated the problem of options for Soviet politicians. But Syria seems for them a fundamental strategic area, and this potential encirclement of Turkey should not be underrated.

---

[4]Edward N. Luttwak and Robert G. Weinland, *Sea Power in the Mediterranean: Political Utility and Military Constraints* (Beverly Hills, CA: Sage Publications, 1979), p. 12.

[5]Richard B. Remnek, "The Politics of Soviet Access to Naval Support Facilities in the Mediterranean," in Bradford Dismukes and James M. McConnell, eds., *Soviet Naval Diplomacy* (New York: Pergamon, 1979), pp. 357ff.

ITALY AND THE SOVIET UNION

What might the next Soviet objective in the direction of Anatolia be? Most everyone remembers the words of the president of the Italian Republic, Sandro Pertini, about the rash of violence that threatened to destabilize Turkey. At present, the outbreak seems curbed but other problems remain, and the first of these involves the Straits. NATO scholars agree that the Straits seem to be the only Mediterranean zone where the Soviet Union could confront the West with a *fait accompli* before it could respond. This possibility is not to be ignored.[6]

Soviet pressure on the Straits is increasing as it attempts to "stretch" the Montreux Convention to the breaking point. For example, on 18 July 1976 the aircraft carrier *Kiev*, excluded from the Straits, passed through them, while the Soviets wrongly announced that it was an "antisubmarine ship." Turkey and its allies swallowed the bitter pill. Moreover, the fact is that the Straits zone could be overwhelmed so quickly as to make any prospect of a "flexible response" there absurd.

A Soviet attack in the Straits zone could not—it is assumed—expect indifference from Greece. With its island territory stretching into the center of the eastern Mediterranean, Greece constitutes the second rampart to obstruct the way toward the center of the basin. But what would the Greek reaction be? What lies behind the anti-Atlantic and anti-American proclamations of some present-day leaders? And what is the meaning of such episodes as the visit of the Soviet prime minister to Athens, or the mooring authorizations to Soviet naval ships? Is it only a calculated game to get more from NATO, and especially from the United States? Although the complicated and traditional hostility between Greece and Turkey should not be underrated, the common danger should limit the shrill rhetoric. It is noteworthy that the Papandreou government has not withdrawn Greece from NATO.

The third rampart is created by the Italian peninsula, the last wall protecting the western Mediterranean and, at the same time, the very heart of Europe. It deserves more extensive consideration also because it constitutes, at least until now, the largest power that is exclusively Mediterranean. One of the main strategic characteristics of the Italian peninsula, with its large islands, is its length. It extends for more than 10° latitude, from 47° 5′30" to 36° 38′ 35". Its two crucial points are, in the north, the Po Valley, access to which is through the Venetian plains and the Isonzo Valley, and in the south, Sicily, whose possession is fundamental for the control of the central Mediterranean. Friedrich Engels wrote in 1858: "More than Flanders, Northern Italy has been for centuries the battlefield on which Germans and Frenchmen fought their wars. The possession of Flanders and the Po Valley is the

---

[6]Gürkan, *NATO, Turkey and the Southern Flank*, p. 48.

necessary precondition for the attacker. . . ."[7] With that he underlined the strategic importance of northern Italy for every battle between nations whose armies advance toward battlefields through the Elbe, Rhine, and Flanders. In other words, "in Western Europe there are only two big plains where any historical confrontation between Western and Central powers inevitably occurred: Flanders and the Po Valley."[8]

Centuries of history of war demonstrate how many times victory or defeat in Central Europe has been decided by the possession of the Po Valley, and how—in one sense or another—it has constantly acted as a strategic throughway to outflank Central Europe. From Turenne to Catinat, from Luxembourg to Prince Eugene of Savoy and Napoleon, all great generals were deeply conscious of this reality. Since geostrategic conditions are unchangeable, the Austro-Hungarian chief of staff, Field Marshal Conrad von Hötzendorf, was wise indeed to plan in 1916 a strategic offensive that would have enabled the imperial and royal divisions to descend into the plains from the Trentino highlands, making the Italian positions on the rivers Isonzo, Tagliamento, and Piave all but useless. This would have led to the collapse of General Luigi Cadorna's armies and to the loss of the Po Valley. The consequences would have been appalling. Italy would possibly have been lost, compelling the Western allies to reach a compromise with Germany. General Eric von Falkenhayn, German chief of staff, did not understand; instead of supporting von Hötzendorf's plan, he chose to push his divisions into the meat grinder of Verdun.

THE YUGOSLAV ISSUE

What lesson did the strategists of the Warsaw Pact learn from this experience? The organization of a Hungarian armored army a few years ago may be a clue. It is difficult to see where such a large unit could be otherwise efficiently employed, except as a part of an invading force attempting to pass from Yugoslavia and the Gorizia gap.

All of this has certainly been taken into account by Italian strategists, but has it also been considered by the highest NATO military authorities? If so, perhaps we can be satisfied about the defense of the crucial border on the Isonzo. What is uncertain, instead, is what Yugoslavia would do on the other side of the border. It seems that a reasonable optimism concerning Yugoslavia is justified, maybe more than that expressed by other distinguished speakers.

---

[7]Raimondo Luraghi, "Il Saliente trentino nel quadro strategico della prima guerra mondiale," in Sergio Benvenuti, ed., *La Prima Guerra mondiale e il Trentino* (Rovereto: Comprensorio della Vallagarina, 1980), p. 13.

[8]Ibid.

Its differences with the Soviet Union are serious, and it is difficult to see how they could be eliminated under the present circumstances; they have deep roots that go back to the Resistance and before.

It seems that Tito, even in the 1930s, had few illusions about the methods and objectives of Soviet leaders. Therefore, despite Yugoslav hints of relaxation toward the USSR, and despite the authorization granted to the Soviet navy to use Dalmatian naval facilities (for which the Soviets pay), there is no reason to doubt Yugoslavia's concern for its own independence. These facilities are used principally as ports for repair and resupply. There is no evidence at this time that the Soviets have built any military installations ashore, or that they possess restricted areas controlled by their own police such as those in Aden. Still, the Yugoslav attitude toward the naval policy of NATO has deteriorated recently. Although the Sixth Fleet was once regarded as the protector of Yugoslav independence, since 1960 the Tito government and its successor have labeled it a threat to Yugoslavia as well as to the Soviet bloc and the "progressive" Arab states.

In the early 1970s Belgrade was upset by the use of Piraeus as a "permanent" base by the Sixth Fleet. Yugoslavia was alarmed also by Italian-American naval maneuvers in the Adriatic in defense of the status quo on the Italian-Yugoslav frontier, notably exercise "Dark Image" which took place in early 1974.[9] Despite all this, relations between Italy and Yugoslavia have remained excellent, and it may be that the pro-Soviet hints in Belgrade are an attempt to tranquilize the potential enemy, an attempt perhaps motivated by the sad fact that little help for Yugoslavia could be expected from the West in case of a Soviet attack.

The major concern of the Yugoslavs is undoubtedly Bulgaria. This country, which is relatively small, has always been a bitter enemy of the Serbs and is today a stronghold of the Warsaw Pact. Relations with Romania are anything but ideal and have even motivated the Soviet Union to build large hydrofoils for the rapid transportation of armored forces to Bulgaria where they could threaten the Turkish Straits, Greece, and Yugoslavia. To anyone who has traveled through the mountainous Iron Gates, it is clear why the invasion of Yugoslavia through Romania is not too probable. The historic invasion route is from Bulgarian Macedonia into Yugoslav Macedonia, and this was used in 1941.

If there is anything that Italy should absolutely avoid that would require NATO to adopt extreme measures, it is the prospect of seeing Soviet and Warsaw Pact tanks in the streets of Nuova Gorizia. Such a danger could materialize in a situation in which a Yugoslav faction, panicked by an economic crisis, could decide to play the Soviet card and request support of the Warsaw Pact.

---

[9]Remnek, "Politics of Soviet Access," pp. 383ff.

THE SOUTHERN LITTORAL

Until a few years ago the Northeast Frontier was the principal problem with which Italy was concerned inside NATO, but developments in North Africa have created another danger. This threat from Libya, in a sense, has outflanked the eastern Mediterranean. As long as Tunisia is hostile to Soviet expansion, the Sicilian Channel will be protected; nevertheless, the massive presence of Soviet weapons in Libya represents, without doubt, the most serious threat to the Mediterranean that has developed in recent years. Regarding the attitudes of the Libyan regime, the Italian government has been vacillating and uncertain, even when the *vox populi* has accused Libya of most of the responsibility for the attempts at destabilization that have taken place in Italy.

Because of the unpredictable nature of its leader, Libya's policy is difficult to discern. It could be that the principal objective at this time is more in the direction of Egypt and Sudan than in the central Mediterranean. That is all the more likely because those two countries are threatened from the south by another client regime of the Soviets—Ethiopia. This could bring back to life Mussolini's plan to crush Egypt and Sudan in a Libyan-Ethiopian pincer. Although merely speculation, it is certain that the collapse of Egypt and Sudan would be greeted in Moscow as a great victory, as it would not only liquidate two of the best friends of the West in the Arab world but also lead to the loss of the Suez Canal and deal a deathblow to other pro-Western Arabs. For this reason, it will be necessary in the coming years to watch the area closely. An effort already was made to carry out this plan with the assassination of President Anwar Sadat, which demonstrated that there are forces ready to do anything to change current Egyptian and Sudanese policy.

The presence of Italy in the center of the Mediterranean implies strategic responsibilities that cannot be escaped. Recently these responsibilities have been assumed by Italy in a much more determined manner than in the past. The decision to install cruise missiles in Comiso is an example. However, it is to be expected that pressure on Italy in this case will increase. The adversaries probably have not yet played all their cards, and a further action, financed with seemingly unlimited funds, is probable.

Compared with the eastern Mediterranean, the western Mediterranean is a secure and safe possession for NATO. This does not mean there are no problems. France, even if it is a valued collaborator, especially with its present government, is not a part of NATO. Spain is marking time so that there is no discussion of integration of its forces. Algeria gives the Soviets the use of ports, but not the base of Mers-el-Kebir. The situation in the western Mediterranean is thus anything but settled, and the most important duties there have to be carried out by Italian and French naval forces.

It is important to emphasize that the armed forces of other Mediterranean NATO powers have at least as many problems. The Turks are in a worse situation. Their tanks, airplanes, and ships are obsolescent, and, if Turkish soldiers are excellent fighters, it is difficult to see how they can fight to full advantage without a radical renewal of their material. The Greek armed forces are in the same situation, with the exception of their air force, while the Spanish armed forces—not yet integrated into NATO—have an excellent but small navy, as well as land forces with out-of-date equipment.

Italy is better off, at least in appearances, and this seems natural for a country that is one of the most industrialized and by far the most important country in the Mediterranean. This does not mean it will not have serious problems. However, when the air force is equipped with the new multirole *Tornado* aircraft, when the armored units are equipped with the *Leopard II* tank, when the artillery will be modernized, and when the modernization of the navy is completed, the situation will improve greatly. Attainment of these goals is not far off.

However, the principal problem for Italy is the numerical reduction of its forces and the insufficient training that all but certain elite units receive. There is a continuous and persistent campaign conducted against the armed forces (rarely in the open) to reduce their numbers, diminish their equipment, and to persuade the public that money spent for defense is "wasted." To paraphrase Raymond Aron, the issue facing the Italians, like the French, is clear: can we remain Italy or must we become Poland? This problem for the Italians is indeed more important than overpopulation in Bangladesh.[10]

CONCLUSION

At this point it is necessary to think of the future. I cannot help but agree with Professor Joaquin Oltra that it is not the business of the scholar to be a prophet. All one can do is to express concerns and ask questions. Perhaps such stimulus will move political leaders to make the right decisions in a rapid and courageous manner in order to answer satisfactorily all the pertinent questions that come to mind. Have plans been made to defeat the poisonous campaign against the armed forces in Italy and to help Turkey modernize its weapons in a rapid and adequate manner? What will NATO do if Libya and Ethiopia begin an aggressive action against Egypt and Sudan? Will NATO restrict its answer to saying that it is not NATO's business? Will it let the Americans, as always, pull everybody's chestnuts out of the fire and then criticize their actions? What will the alliance do in Yugoslavia if, after a political struggle, a leader should emerge who would request "fraternal assistance" by the Soviet Union? What will it do in the event of a sudden

[10]Raymond Aron, *Le spectateur engagé* (Paris: Julliard, 1982). Italian translation by Alberto Ronchey, in *L'etica della liberta, memorie di mezzo secolo* (Milano: Mondadori, 1982), p. 274.

strike against the Turkish Straits, or if the Iranian chaos produces a pro-Soviet government that would have the effect of bringing Soviet forces to the Persian Gulf? Will NATO continue to say that it is not involved? Could Italian, French, and Spanish naval forces agree on a common policy in time to permit the United States to send part of its Sixth Fleet there?

However, there does not appear to be any sign of a response, at least on a political level. NATO is far from real coordination of naval forces in the Mediterranean which would permit the flexibility mentioned above, and the danger of a possible deterioration of the situation in Yugoslavia does not seem to be taken seriously. It is useless to refuse to look at the future of NATO in the Mediterranean. The future of the West and that of its children and their right to grow and live as free people depends on answers to these questions.

THE BALKAN POWERS

Yugoslavia: The Balkans, the Mediterranean, and the West

STEVAN K. PAVLOWITCH

General John Hackett's "Third World War" starts in Yugoslavia in July 1985.[1] From time to time the West becomes nervous about that country. For example, in 1968, after the Soviet intervention in Czechoslovakia, NATO had sent signals to deter the USSR from any move against it. Then at the beginning of 1981, when Soviet armor had moved into Afghanistan and Tito lay dying, Western eyes were again turned on Yugoslavia. Suddenly, the land of the South Slavs was no longer the dashing and progressive leader of the Third World but a poor and unsteady country on the other side of the Adriatic that could become a future Soviet target. News commentators said that, if the Russians were tempted into what in 1968 had been called the "grey zone of NATO," the whole strategy of the Atlantic Alliance in the Mediterranean would have to be reconsidered. The crystal ball gazers of political science and the players of international relations party games already had been indulging for a number of years in such speculation and writing scenarios of intervention. They generally agreed that the great attraction for the USSR in seeking to obtain control of Yugoslavia was access to that country's coastline.

What could NATO do to counteract a direct Soviet intervention? Only an armed attack on one of its members binds the Alliance to assistance, and there is no legal basis upon which the organization as such could operate militarily on the side of Yugoslavia. At the same time, the total security requirements of the alliance call for a broader view. In spite of a reluctance to grapple with "out-of-area" problems, it is now widely recognized that such requirements are directly affected by events beyond the NATO area. The allies could decide, individually or collectively, to extend material aid to a universally recognized government against unprovoked aggression in answer to a request by that government. But an analysis of the forces available for the defense of the alliance in Europe, and of the ways in which such aid

[1]To be precise, the fictional preliminaries start in Yugoslavia with a Soviet incursion into Slovenia. According to Sir John's scenario, the Soviet-sponsored Committee for the Defense of Yugoslavia, set up in the northwest of the country, has called for help from the USSR. Within days, Soviet units were in action against U.S. troops that had arrived from Italy. The Soviet move was nicely calculated, says the author: if the West did nothing, it would be a quick and easy gain and a warning to the Warsaw Pact allies; if the West did oppose the move with force, this would be an attack on a peace-loving Socialist country such as to justify defensive action against NATO. John Hackett, *The Third World War: The Untold Story* (London: Macmillan and Company, 1982), pp. 2–4.

could be brought to Yugoslavia, reduces the probability of direct assistance rapid or massive enough to be effective.[2]

THE ORIGINS OF YUGOSLAV NONALIGNMENT

The nearest Yugoslavia ever came to NATO was in 1953–54 when, in the wake of the Korean War, a frightened Tito reluctantly committed himself in some manner to the Western alliance in return for protection, by associating Greece and Turkey in the Balkan Pact. In Yugoslav eyes, however, this was only a temporary arrangement, and within a year of Stalin's death, the pact already had become a dead issue.

Yugoslavia's foreign policy then became a blend of two ingredients, one emerging from Tito's longing for a bigger role, and the other from his country's geopolitical position. Whereas the former was revolutionary, ideological, and loudly proclaimed, the latter was traditional, pragmatic, and unacknowledged. The area that makes up Yugoslavia always has been one of convergence and transition; nonalignment is only the last in a long series of historical situations where Yugoslavia or its components have found themselves in the position of neutrals or intermediaries. In modern times, Yugoslav leaders have acquired something of a reputation as tightrope walkers or jugglers. They have established a tradition of balance, which also is one of ambivalence, between Russia and Europe. Whenever the international order has appeared favorable to Yugoslavia, their diplomatic policy has been to try hard to consolidate it.

Tito obviously would not have acknowledged such an ancestry. Right from the start the foreign policy of the Yugoslav Communists looked out ambitiously beyond their Communist neighborhood to the Mediterranean and to the entire continent of Europe.[3] There is, however, no real evidence of

---

[2]In preparing this paper the author naturally has turned to some of his previous writings: "The History Behind Yugoslavia's Psychological Deterrent," *Review of the Study Centre for Yugoslav Affairs* 2, no. 2 (1975): 107–21; "Où va le titoisme?" *Commentaire* 2, no. 5 (Spring 1979): 93–100; "Yugoslavia After Tito: The Grey Zone on NATO's Balkan Flank," *Survey* 25, no. 3 (Summer 1981): 20–42: "Le titoisme et les constantes politiques yougoslaves," *Commentaire* 4, no. 16 (Winter 1981): 533–41; and "Kosovo: An Analysis of Yugoslavia's Albanian Problem," *Conflict Studies*, no. 137–38 (1982): 7–21.

[3]As early as 2 May 1944, Edvard Kardelj was writing to Dedijer, with praise from Tito: "We must already now strive for our country to become an influential factor in the setting up of the future international order." So as to come to the conference table with a strong international position, Yugoslavia was to establish links that went further than the Communist countries, to the "freedom-loving forces" of the Mediterranean, France, Greece, and Italy. Five months later, in his first speech in liberated Belgrade on 27 October 1944, Tito said: "We have acquired the right to take our place as equals with the Allies in the war, and also in the building of a new and happier Europe." Vladimir Dedijer, *Novi prilozi za biografiju J. B. Tita*, vol. 2 (Rijeka: Liburnija, 1981), pp. 883, 920.

their "struggle against spheres of interest" going back to 1943, as Vladimir Dedijer would have us believe, just as it does not make sense to claim that at the time they were opposed to the provisions of the Yalta Conference and the arrangements that had preceded it.[4] At Tehran, Tito was spared the possibility of a landing in Dalmatia which he feared so much, and it was agreed there to help his movement to the greatest possible extent. By the time of the Churchill-Stalin meeting in Moscow, the Red Army was conducting joint operations with the partisans in Yugoslavia at the request of Tito, who already held most of the country. At worst, the "fifty-fifty" deal, which had amended Molotov's "sixty-forty" proposal, did Tito no harm; at best, it helped him. The rough equilibrium envisaged by Churchill was later to become the bulwark of Tito's nonalignment. As for Yalta, it gave him the opportunity to complete and formalize his control over Yugoslavia.

The Yugoslav Communists had successfully turned a foreign war into a revolutionary struggle and brought the country solidly into the Soviet camp. They were full of self-confidence, and their foreign policy, like their domestic policy, was driven by revolutionary dynamism. If the new regime's policy toward Italy and Austria could be considered as aiming to complete the unification of the country's ethnic territory, its policy toward the southeast was truly expansionist. By the end of 1947, the dream of a Yugoslav-led Balkan federation seemed to be taking shape. At a time when Stalin still had to tolerate a good deal of diversity in Eastern Europe, the real differences between the Yugoslav regime and the regimes of the other Communist countries were the product of war and geography. They lay in the extent of Tito's hold over his party and his state, and in his country's position bordering on the Mediterranean and on non-Communist Europe.

The crisis eventually arose from Tito's extra-Yugoslav ambitions. It was difficult enough to keep any control over his domestic establishment, but this was nothing compared to the liability that the Kremlin would have to face if Tito's government succeeded in creating a vast complex expecting to share in the formulation of overall Communist-bloc policy, and that with a naiveté which disregarded the realities of power at the global level. Because of increasing tension in Europe, the Soviets wanted to prevent the development of a situation likely to cause trouble in the Communist camp itself. They wanted to be able to control Yugoslavia more directly in view of its important strategic position.

It was his instinct of self-preservation, along with his love of power, that made Tito resist the Soviet leadership. This, together with his desire to

---

[4]Dedijer, *Novi prilozi,* passim, esp. pp. 749, 811, 853, 862, 914. His hypothesis of the "widest coalition in the Second World War"—British plans to liquidate the Yugoslav Communist movement before the arrival of the Red Army, through the partisans' native rivals, with the connivance of the Axis and the happy abstention of the Russians—points less to Anglo-American intentions than to the partisans' obsessional fears of an Anglo-American landing and to their distrust of the Western support that had helped them through a stage in their struggle when their survival was still at stake.

spread communism, led him eventually to support almost any take-over bid elsewhere by a Communist, Socialist, or "progressive" minority, provided it was not done with direct Soviet military involvement. In the initial period of indecision following the break with Stalin, Tito had sought to find his bearings while maintaining an equidistant position between East and West. This made him more sensitive to the changing balance of world forces. In the mid-1950s the Yugoslavs turned toward developing the contacts they had made earlier on the Indian subcontinent, when they were playing an important role in the Cominform, and then broadened these initial successes to ever-expanding contacts with the newly emerging nations of the postcolonial world. The rallying call came to be known as "nonalignment." Yugoslavia's ties with the Third World allowed it to weaken its links with the West and facilitated reconciliation with the Soviet bloc on its own terms after Stalin's death. Over the 1950s and 1960s Yugoslavia's leaders came to envisage an ideal "socialist" world of equal "Socialist" forces, with an important role for themselves in promoting "socialism."

By the late 1950s they had concentrated their attention on Nasser's Egypt and from there branched out into the Arab world and the Algerian independence movement. Of the three centers of revolution and socialism vying for influence in Asia and Africa, Belgrade, with its concern for national independence, probably was closest to the feelings of Afro-Asian nationalism. Tito increasingly made use of international opinion outside the blocs to appear as a world leader. The selection of Belgrade as the site of the first nonaligned summit conference in 1961 was Yugoslavia's greatest diplomatic success, but, as early as that meeting, its nonalignment showed an anti-Western bias. The second conference, held at Cairo in 1964, already reflected the drawbacks of nonalignment as a united movement. Relations with France in particular deteriorated seriously over Tito's aid to the Algerian FLN, and the West in general was reminded of the quintessential fact of Yugoslavia's communism by the apparently narrowed gap between the foreign policies of Belgrade and Moscow. While determined to preserve its independence, the former shared with the latter the Communist view on a whole range of questions.

Yugoslavia's prestige was highest in the Arab Middle East where it backed Nasser's pan-Arabism. Tito played a perceptible role during the 1967 conflict. From the start, he had given Egypt complete support. Nasser's defeat came as a blow, for without him Yugoslav influence would be greatly reduced. Tito did everything in his power to help restore Nasser's military fortunes, by fully cooperating with and urging on the Soviet bloc. More realistic counsels eventually having prevailed, Tito turned toward mobilizing world opinion in order to restore, as much as possible, the prewar situation. His venture, approved by the Warsaw Pact allies, was brushed aside by Israel and was not followed with enthusiasm by the Arabs either.

The Yugoslavs also tried to promote nonaligned socialism on a Mediterranean scale through conferences of "progressive" political organizations.

They denounced the presence of all nonriverain military forces, but at their 1968 Rome meeting, the progressives generally preferred to limit their final declaration to the exclusion of the United States and the United Kingdom, and Yugoslavia did not sign. It took its stand on the belief that the idea of socialism was far stronger in the Mediterranean than was realized by the Soviets, whose military presence, far from being essential for its promotion, could actually compromise it. Its position was, nevertheless, basically anti-American; if the Mediterranean could be rid of U.S. forces, there would be no need for Soviet forces to balance them, and socialism in the area would be able to direct its own destiny. Among other things, Yugoslavia's missionary zeal certainly helped to create an image of the USSR as an outsider in the Mediterranean.

Such was Tito's way of dealing with the Soviets. When they were cordial, he was only too happy to be friendly. When they tried pressure, he resisted and made sure that the world knew about it. Yugoslavia's foreign policy in the 1960s was on occasion as close to that of the Soviet Union as the stand of any Warsaw Pact ally because Soviet interests at times coincided with those of Yugoslavia. If Tito generally aligned with Moscow in defense of the Arab cause, it was because he was no neutral in the Middle East. The Arab cause was one of the vehicles for Tito's promotion of socialism in the Mediterranean. Independent though it was, Yugoslav diplomacy was that of a party leadership anxious to play a leading role in the international movement gathered around the concept of Communist socialism.

While stressing the principle of equality between Communist parties, Yugoslavia was finding more and more in common with the USSR as the 1960s went on. Paradoxically, when Yugoslavia's vital interests appeared to be placed in jeopardy by the 1968 Soviet intervention in Czechoslovakia, it was, in fact, moving back to its previously close relationship to the USSR. Having institutionalized nonalignment as Yugoslavia's constitutional foreign policy, Tito had realized that the influence of the nonaligned movement in world affairs was fast declining.

Imposed originally by circumstances, the Yugoslav policy of nonalignment had reached the stage where ideology had almost completely displaced all considerations of geography, history, economics, or culture. The country's rulers had come to look upon themselves as leaders of the Third World and as the main propagators of socialism there. This international role brought Yugoslavia a measure of independence, with much glamour. Tito's influence in Asia and Africa was probably one of the reasons for the attention paid to him by the USSR. It satisfied the vanity of the Yugoslav leadership and ended by feeding Tito's narcissism. At the same time, it has to be said that the Third World, while admiring Yugoslavia's ability to maneuver between the superpowers, never paid too much attention to the actual nature of its domestic policies. It also could be argued that Yugoslavia had helped to spread Soviet influence among the nonaligned countries of the Mediterranean and Middle East, thus in a sense endangering its own position.

YUGOSLAV FOREIGN POLICY IN THE 1970s

Because of his near veneration of Soviet Communist ideology, Tito often misread the Kremlin's intentions, as in the case of the Czechoslovak crisis which, for a while, again worsened relations with the USSR. Yugoslavia went through a period of turmoil after 1967. Tito was trying to be the guide of a movement of uncommitted states; his country was looking westward, culturally and economically, and yet ideologically it was tied to the East. It felt in danger as a result of Soviet action in Czechoslovakia, and yet the uncommitted did not raise their voices. Yugoslav nonalignment was turned into an "antibloc" diplomacy. The emphasis changed to "defending the country's independence and integrity." The rhetoric was that the Yugoslav system offered the only genuine opposition to the two established power blocs, each with its own sphere of influence, neither of which had been able to improve the condition of contemporary man.

It was in 1968 that the Yugoslavs discovered their opposition to the Yalta agreement. The idea had been taken over from Gaullist France where Yalta had become one of the symbols of the division of Europe, but it suited Yugoslavia's new stance. It also helped to curry favor with General de Gaulle; Tito was anxious for French support, which was less embarrassing than any American connection. Broadly speaking, more attention was paid to Europe, not only to France but also to Italy and eventually all countries of the European Community. This meant welcoming, and even attempting to lead, a loose gathering of southern European Communist parties in opposition to Soviet attempts to reestablish authority over the world Communist movement. At the same time, the Yugoslavs had to be careful to balance this advantage against the detrimental effects that the Euro-Communists' pluralistic approach might have on party authority at home.

Conditions in the Balkans favored the renewal of proposals for regional cooperation. These had first been produced in 1957 in the context of a series of schemes inspired by Krushchev for the reduction of East-West tension through the weakening of NATO. The initiative now came in a different context from the Greek government of Constantine Karamanlis and led to the conference held at Athens in 1976, attended by Bulgarian, Greek, Romanian, Turkish, and Yugoslav representatives. This meeting produced a long list of suggestions for cooperation in nonpolitical areas. Yugoslavia, which had welcomed the effort, nevertheless showed considerable skepticism, reflecting fear of great-power manipulation, particularly of outstanding bilateral issues such as Macedonia, and the formal need to stress its nonalignment.[5]

---

[5]Yugoslav officials maintained that their country could have no special Balkan policy because it had a global concept of nonalignment. As one diplomat told an American analyst, Yugoslavia "has emerged from the narrow Balkan context into the world scene as an active participant in global affairs." Andrew Borowiec, *Yugoslavia After Tito* (New York: Praeger, 1977), p. 70.

Additionally, Tito traveled to both Greece and Turkey in 1976, offering to mediate and generally trying to exploit both countries' disenchantment with NATO, for the vague purpose of creating a nonaligned Balkan zone. While this can be dismissed as empty rhetoric, the presence of Markos Vafiades in Yugoslavia in the summer of 1978 is surely indicative of a specific Yugoslav Balkan policy. Vafiades, who has since returned to Greece, had been closely connected with Yugoslavia's support for the earlier Communist uprising in Greece. Could he have come on a holiday to Skopje, from the USSR where he had been living since 1950, without the approval of both the Soviet and the Yugoslav authorities? In 1982 the major Yugoslav daily paper serialized extracts from a forthcoming book by a Yugoslav Macedonian journalist who described the author's encounters with "General" Markos in 1978. Was it to show that the Yugoslavs had not really stabbed the Greek patriots in the back at the end of the civil war, to put their record straight with the new Socialist administration in Athens, perhaps even to find favor in the eyes of the Greek Communists?[6]

Throughout the 1970s there also was a process of "normalization" with Albania. One major obstacle was the presence in Yugoslavia of a large Albanian minority living mostly in the border province of Kosovo. Tito's strategy was to enmesh the Yugoslav Albanians into his system by giving them all the cultural rights they could dream of in return for continued commitment to Yugoslavia and its regime. He might have thought that through normalization Yugoslav ideas might find their way across to Albania through Kosovo. If so, and had he lived a little longer, he would have been disappointed to see that it was Albania that had infected Kosovo rather than the other way round.

Yugoslavia's independent status is the result of its geopolitical situation as the Communist-ruled territory most exposed to the West, of its own Communist movement which arose out of the civil strife in the Second World War, and of the exceptional amount of help it received from the West. Even the internal political consolidation that took place during the 1970s was backed indirectly by Western financial assistance. Yet such support could never be sought publicly or even acknowledged for fear of the consequences that a westward turn would have on the Communists' domestic monopoly of power, even though the main threat to the country's security since 1948 has come from the East. The antibloc policy assumes the two blocs to be similar. Officially, it cannot see that the West hardly forms a homogeneous unit in its reactions to Soviet imperialism, let alone on other issues, and that even the United States usually pursues a policy of opposition to the extension of

---

[6]Markos's stay in Yugoslavia was first revealed in an interview with the popular Belgrade newspaper *Večernje novine,* 28 November 1979. The extracts from Joran Popovski's "General Markos: Why Stalin did not shoot me" were published in *Politika*, 3–16 June 1982. Markos returned to Greece a few days after Prime Minister Nicolay Tikhonov's visit to Athens in February 1983, under the terms of measures introduced by the Papandreou government to bring the civil war to a final close.

Soviet influence rather than an anti-Communist or even an anti-Soviet policy as such. Yugoslavia still upholds a view of the United States as the enemy of socialism and of political and social progress generally.

Foreign policy was part of Tito's personal prerogative, and he saw himself not as the embodiment of defiance to the Soviet Union but as one of the world leaders of communism. He considered that the success of the Yugoslav revolution entitled him to that status. He believed that his tactics were subtler than those of the Soviet leadership. He wanted to build himself into a world figure, a role for which his country alone did not provide a big enough stage. Since the Communist camp would not allow him to play this role, his government devoted enormous efforts to building up a progressive movement of nonaligned countries which he would lead. All the time, however, he retained a nostalgia for the Socialist world, cleverly used the West to protect him against the Kremlin, and did all he could to figure prominently on the Communist front, always attempting to have the best of all worlds. Nevertheless, in the role that Tito played his only real base was Yugoslavia, his initial catapult but also his limiting factor. An independent Communist and European country, Yugoslavia posed as part of the developing world and kept an allegiance to a nonalignment that had become a mythical monster and a dogma.

Tito's arrogance, which consisted of making moral judgments on every issue under the sun but in accepting only praise from the outside world, was part of his policy of grandeur. He assumed that he could remain safe in every crisis by assigning blame evenly to both power blocs, from behind the facade of a nonaligned structure where Yugoslavia's influence had diminished and whose style had become too radical for Tito's liking. While this policy still gratified his vanity, it no longer offered the Yugoslav leadership more than a convenient way of slipping out of certain obligations toward the Socialist states of Eastern Europe and of avoiding the consequences of its ties with Western Europe.

Having put his own house in order in the early 1970s, Tito, in any event, had resumed his close relationship with the USSR. Ideological in its essence, in its ambitions, and in its appearance, Yugoslavia's foreign policy has been pragmatic only in its tactics. Tito had been careful to combat the impression that he was inclined to appease Moscow. He always had managed to convey to the West his anxieties about Soviet pressures, as well as his willingness to resist them, provided the requisite degree of Western backing was forthcoming. Once he had obtained the right response, he was quick to stress how good Soviet-Yugoslav relations were. The same behavior was forthcoming from his heirs at the time of his death.

The Yugoslav leadership has nurtured a sense of chronic threat to the country's independence, not only from the East where it has some reality but also from the West where Yugoslavia really has few enemies. The behavior of Tito's regime in his last years—repression of potentially anti-Soviet or pro-European elements, gratuitous display of an anti-Western stance, and significant concessions to the USSR—was creating an atmosphere, at least in the

United States, which was not very conducive to Western support of an imperiled post-Tito leadership. Was Tito thereby trying to diminish the Kremlin's incentive to resort to armed force against his successors?[7]

During thirty-odd years Yugoslavia had tried to implement a global foreign policy. Tito's costly diplomacy, up to a point, had covered his country's internal problems, but it had failed to solve them. These problems are now too big to be concealed any longer, and Tito is no longer there to want to do it.

POST-TITO FOREIGN POLICY

An uncertain or chaotic situation could turn Yugoslavia into a Soviet target, but Soviet leaders considering an occupation of that country would have to weigh a number of factors such as the degree of resistance they would encounter, the need to claim that they were answering a call for help, the extent of the reliance they could place on the Warsaw Pact allies, and the nature of any reaction from the West. A direct intervention is unlikely except as a last resort in an internal or international crisis. Even then, would the Russians risk a world crisis for a country that may well be, in global strategic terms and in a perspective that has become fluid, less important than it is made to appear?

The Soviet leaders have come to accept that Yugoslavia is a special case. They know that its independence is considered by the West to be an important factor in international relations. They cannot afford to cope with Afghanistan, Poland, and Yugoslavia. Even Soviet hawks would agree that the best policy is to let Yugoslavia become weaker. Closer economic ties having been achieved, pressure could increase to obtain political and strategic concessions.[8] Soviet pressure could be exerted through separatists and through

---

[7]In the spring of 1975, Tito had told a Czechoslovak military delegation: "Formally we are not members of the Warsaw Pact, but if the cause of socialism, of communism, of the working class, should be endangered, we know where we stand. . . . We hold our aims in common with the Soviet Union." *Narodna armija*, 1 May 1975. By the end of 1975, Yugoslavia's ruling League of Communists (LCY) had put it in writing: nonalignment was an anti-imperialist policy with a class content leading to socialism. *Kommunist* (the LCY weekly), 15 December 1975. The following year the Conference of European Communist Parties held in East Berlin in June was able to see the LCY as a "progressive" force.

[8]In 1973 Rijeka was used as an embarkation point for Soviet heavy equipment going via Hungary to the Middle East. Since then, the Soviet navy has increased quietly its utilization of Yugoslav ports, with supply and repair facilities, and its own floating dry dock in the Bay of Kotor. The Soviet Union has intermittently been granted overflight privileges by Yugoslavia for various flights, including those involved in logistic support to Egypt and Syria in 1973, similar support to its clients in Angola in 1976, and ongoing Soviet reconnaissance flights on their way to the Mediterranean. Gavriel D. Ra'anan, *Yugoslavia After Tito: Scenarios and Implications* (Boulder, CO: Westview, 1977), p. 109; Shelford Bidwell, "Current Objectives and Soviet Strategy," in *The Soviet War Machine*, ed. Ray Bonds (London: Salamander Books, 1976), p. 48.

neighbors. Threatened with dismemberment, and having lost much freedom to maneuver, Yugoslavia's leadership well might want to return to the Soviet fold and claim that Titoism had thereby been saved. The country actually could slide back into place without much drama or too many international complications. The leadership of the CPY, finding itself in a dilemma after 1948, managed to become free of its loyalty to the USSR. Why should not the leadership of the LCY (the new name for the CPY, or League of Communists) be able to drop nonalignment and return to the fold? The Soviets have been very active in Yugoslavia since Tito's death. Soviet interest in the Adriatic coast has been sharpened by an appreciation of the economic resources of its hinterland, while at the same time the USSR is the main supplier of oil, gas, and heavy military equipment. In fact, Yugoslav trade with the Soviet bloc has now reached a level that gives Moscow considerable political and economic leverage.

For a year or two after his death, Tito's successors tried to pursue two aims: at home, to check an ever more worrisome economic situation; and abroad, to maintain some semblance of a nonaligned movement. They placed their undertaking under the sign of Tito's continued posthumous omnipresence. The cult grew into unbelievable proportions, inspired unwittingly by a pseudo-Christian conception of life after death, with echoes of the real presence, incarnation, and transfiguration. Tito's heirs continued to move around the globe, trying to keep on schedule the preparation of the nonaligned conference which was to have been held in Baghdad in 1982, attempting to mediate between Iraq and Iran, discussing the situation in southern Africa, proclaiming their support of the PLO and calling for a Palestinian state, condemning Israel's "genocide" in Lebanon, and demanding its withdrawal from all territories occupied since 1967. At the 12th Congress of the LCY in June 1982, they gave a prominent place to delegations from the PLO, SWAPO, and the Communist parties of Italy, China, France, Greece, and the USSR. They did their best to stir up ill feelings between Athens and Tirana (after Albania's overtures to Greece) and between Ankara and Sofia (over the Turkish minority in Bulgaria). And they attacked Western governments for allowing Yugoslav opposition groups.

POST-TITO DOMESTIC POLICY

Inevitably, all this bustle about world policy, particularly the more distant problems, was now in a lower key. It no longer had Tito to act as a resonance chamber, it sounded less convincing, and it raised no interest whatsoever at home. Furthermore, the need for austerity had forced the government to reduce the cost of its sumptuous diplomacy and to cut its vast network of missions. By the beginning of 1981 inflation, running at over 50 percent, was seriously affecting living standards. The trade deficit had risen to the point where the servicing of the foreign debt was approaching the "red zone"—25 percent of export earnings. By the end of the year hard currency obligations were running at $20 billion—almost $1,000 a head and 30 percent of the GNP.[9] The rising price of oil had exacerbated but not caused the imbalance. Investments had systematically exceeded real savings; they were financed by domestic credits at negative interest rates and sustained by imports purchased with hard currency foreign credits.

Yugoslavia was anxious to stress that it was a nonaligned country with good borrowing habits, in whose credit worthiness the West had an obvious interest. No sooner had Tito been laid to rest in 1980, than the government introduced a stabilization policy including a 30 percent devaluation, compressed domestic demand and wages, reduced investments, and channeled resources into exports to the West. By the end of 1981, however, in spite of further drastic import reductions, coupled with a crash increase in exports just to stay out of the red zone, it was clear that the stabilization policy had failed to come to grips with the economic crisis. Rampant inflation already had offset the effects of devaluation and was being caught up again by wages and embezzlements. The effort was, in any case, so heavily subsidized that it could not possibly be sustained for long. It was admitted in ruling circles that the government was unwilling to take such economic measures as might have unpredictable political consequences. It feared a Polish-type crisis. The reluctance of Western creditors to lend more caused a continued squeeze that seemed to subordinate all economic policies to the achievement of balance-of-payments targets.

In June 1982 the government finally admitted publicly the real situation: that Yugoslavia would have to spend more than 50 percent of all its earnings on debt repayment due that year; that inflation could not be down to 15 percent by the end of the year as planned, but that a more modest 25 percent could be attempted if all were aware of the need for discipline and economy; in fact, it turned out to be nearer 35 percent. In short, it was acknowledged that economic recovery could not even begin without a modicum of public confidence.

The deep structural imbalances also had been revealed with unexpected swiftness and intensity by the increasingly voluble self-criticism that began soon after Tito's death, grew during 1981, and had come to question the validity of self-management in its present state—a veil of legitimacy rather than a reality. It denounced the political manipulations that had managed both to make a mockery of the autonomy of self-managing institutions and to break the country's economic unity, the tie-up between the inconsistent economic reform of 1965 and the renewed politicization of the economy after 1971. After nearly three decades of destroying the old social order and replacing it with the zigzag and stop-go "Yugoslav way to socialism," the discrepancy between its theory and its practice has left people confused and vulnerable

---

[9]At that time, Poland's debt was estimated to be $27 billion, or 10 percent of its GNP.

and cadres bewildered and cynical. Yugoslavia's crisis is part of the general East European tension between societies and the regimes that govern them, in spite of all the obvious differences that merely make Yugoslavia a more complex case.

The end of a long reign has released a yearning for change, mixed with apprehension. The most dynamic forces in society are pulling westward; the controlling political force is refusing to go in that direction, but the government has run out of ideas and is unsure of itself. A possible short-term future would be an acceleration of self-repression: Yugoslavia's own rather dangerous mechanism based on inefficient totalitarianism, fear of the Soviet army, fear of internecine conflict, and the feeling that the West is losing interest. New and contradictory developments, however, began to appear in the spring of 1982.[10] The first signs of de-Titoization were in the air.

A quiet de-Titoization is inevitable. To start with, much of what was distinctly Titoist in foreign affairs is already fading. By pointing out some of his mistakes, it might be possible for his successors to shift the burden of blame for present problems back to him.[11] It seems that a substantial number in the LCY would favor an overhaul of the system. There certainly are deep divisions, and a controversy over democracy was not far from erupting within the leadership at the time of the 12th Congress. Congress met in a climate of economic exhaustion. Although the atmosphere was one of pessimism, there was no showdown for the simple reason that fundamental questions were not tackled. Speeches and resolutions repeated descriptive clichés and denounced anonymous culprits.[12] Nobody had the courage to ask for structural changes.

It was revealed in November 1981 that, before his death in 1979, Edvard Kardelj had warned that if Yugoslavia failed to solve its problems quickly it could "experience the fate of Chile." Coming from the weekly *NIN*, a publication under attack from the hard-liners, this could have been an appeal to the party moderates to put their house in order. Tito had designated the army as the bastion, not only of the country's independence but also of its social and political order, and ever since the 1970s, Yugoslavia has seen a creeping influence of the military.[13] There have been angry words spoken by top

---

[10]One way for the Communist leadership of Serbia to voice its concern over developments in Kosovo was to allow the media to report them fully, and spontaneous demonstrations were seen on television for the first time. Since it was difficult to confine that freedom to the Kosovo issue, the Belgrade press became increasingly independent and critical. A number of other important questions about Yugoslavia's present and past were raised, questions too about Tito's legacy and different ways to approach it. Tito was no longer there to impose a consensus.

[11]As suggested by Chris Cviić on the BBC Third Program, 30 May 1982, in his review of Dedijer's *Novi prilozi*.

[12]Jean-Claude Pomonti gave "le recours à la palabre" as a characteristic feature of Yugoslavia after Tito, "La Yougoslavie sans Tito," *Le Monde*, 5 May 1981.

[13]Yugoslavia is merely an instance of a more general trend in Eastern Europe, where the army retains its cohesion in the face of crises in other official institutions. Poland represents the utimate point, when only the army has retained its ability to act. See Jacques Rupnik, "From 'Party State' to 'Army State,'" *The New Republic* (6, 13 January 1982): 17–20.

generals over the last year, implying that the politicians were allowing the situation to get out of hand. Divisions within the leadership are giving the generals the impression that the army may be the only reliable guardian of Tito's legacy. The military is certainly the most consistent, the most conscientious, and the most traditionalist keeper of that inheritance.[14]

The uncertainty that hangs over the future has made the leadership agree on one point: to increase the defense budget even though it is finding it increasingly hard to finance such an effort and the Yugoslav army provides more quantity than quality.[15] Its equipment is aging, standards of maintenance are poor, its backbone remains the traditional infantry battalion, and the training of conscripts is largely a formal exercise. These deficiencies are covered up with the trendy cosmetics of the Territorial, or All-People's, Defense, which exists mainly on paper, in words, and to give veterans the illusion of continuing to serve party and country. The army's increased political role could hamper its capability to defend the national territory against foreign aggression. The doctrine preached to all ranks is that the country is threatened by American imperialism and its pawns (such as Italy and Austria) no less than by Soviet hegemony and its auxiliaries (such as Bulgaria and Hungary): NATO is as dangerous as the Warsaw Pact. If believed, the idea that both blocs threaten Yugoslavia can only cause defeatism. The dissuasive value of Yugoslavia's defenses is obviously less than is made to appear. At best, its military efforts add up to no more than a psychological deterrent of limited power.

THE POST-TITO FUTURE AND THE WEST

Western governments are really interested in only one aspect of the "Yugoslav way," its apparent independence and stability. They are anxious that it should not give way to the threatening powers of the day and fear that centrifugal forces will gain the upper hand as soon as the state is not governed by an authoritarian regime. Their support for Communist Yugoslavia has been anti-Soviet, a support that its leaders have shared with their Chinese opposite numbers as well as with various monarchs, generals, colonels, and other authoritarian rulers over the past thirty years. Repression has generally been ignored and the fear has even been expressed that, without Tito, Yugoslavia might democratize too quickly, almost to the point of provoking Russia. Not

---

[14]To the point where one Yugoslav journalist told the author in the summer of 1982 that the military leader, if ever one emerges, will not be so much a Yugoslav Jaruzelski or a Yugoslav Pinochet as a Yugoslav Tejero.

[15]The Yugoslav arms industry supplies much of the country's needs and has become a thriving exporter. Yugoslavia is one of the top ten arms exporters and one of the biggest suppliers to the Third World: £120 million in the first nine months of 1980. *Times* (London), 16 December 1980. Yet high-technology weaponry and spare parts still have to be imported, and the backbone of its armor and aircraft comes from the USSR.

only does the West see no valid alternative to Titoism, but there also is much unreasonable Western admiration for what was known as "Tito's Yugoslavia" and for the personality of the late Marshal Tito in particular, based in part on the impression that the system he has bequeathed to his country is basically pro-Western.

The cult of Tito has been one of the elements that enabled the Yugoslav leadership to maintain an appearance of continuity and to profit from his prestige in the East where he was known and esteemed, in the West where he was less known and more admired, and among the nonaligned to whom he had extended help and a certain significance in world affairs. Tito's was one of those authoritarian regimes that Western democracies favor so long as they provide apparent stability without too much pricking of liberal consciences. Mediterranean boastfulness with esoteric verbiage works wonders in the West, as Mussolini had discovered long before Tito.[16] Most hues of the Western, particularly the British, political spectrum have found something useful in Titoism at one time or another.[17]

As the price for Yugoslavia's independence from the Soviet bloc, the West also has underwritten the mismanagement of its economy. In the 1970s bankers had been only too happy to lend in order to help recycle the massive surpluses of OPEC countries and hardly needed any political prompting. By the beginning of the 1980s, however, they found themselves in the tricky situation of having to reschedule debts in order to enable nearly defaulting nations to meet interest payments. The Yugoslav government, which is among the sounder of the great borrowers, has managed to meet its 1982 debt schedule with enormous difficulty and has set itself against any rescheduling. It argues that this would damage the image of the country's distinctive brand of regime,

---

[16]A month before Mussolini's declaration of war on Britain, Neville Chamberlain was still expressing to the Italian ambassador his admiration for *Il Duce*, "the greatest man in Europe," as noted by Giuseppe Bottai after seeing Ambassador Giuseppe Bastianini on his return from London. *Diario 1935–1944* (Milan: Rizzoli, 1982), p. 213. More than one year after Tito's death, Secretary of State Alexander Haig wrote in the visitor's book at the late president's tomb: "The great leader and world statesman who has led the Yugoslav peoples out of the ruins of the war, to stability and prosperity at home, to respect and prestige in the world." *Politika*, 4 May 1982. The quotations are translated back, from Italian and Serbo-Croatian, respectively. Verbal inflation has increased between 1940 and 1981 by as much as the world has shrunk.

[17]Great Britain is in the forefront of the Tito cult (for example, the Marshal Tito Fellowship at the School of Slavonic and East European Studies of London University, Malcolm Williamson's "Ode to Tito," and Sir Fitzroy Maclean's consultancy for an epic film on Tito's life and work), so much so that it will probably survive there long after it has faded away in Yugoslavia. There was no continuity between British support of Tito in 1943–44 and the post–1948 support, but the events of 1948 came as an unexpected gift of history to present a coherent story of a thriving, continuous friendship between Britain and Communist Yugoslavia. The self-publicity of British friends of Tito's Yugoslavia has contributed to this in no little way, with the war tales of participants who have naturally lined up their memories with what has happened since. Even though the automatic supporters of Titoist Yugoslavia are now retiring, the general tendency of Foreign Office diplomacy, which equates each nation with its current government and sees to it that foreign rulers be not upset, will help to sustain these ideas beyond their natural life.

system, and policies. Yet the rescue operation mounted to enable Yugoslavia to meet its 1983 schedule is little more than a face-saving rescheduling. Western bankers have come to see their dilemma, caught as they are between having to exert more sway over Yugoslavia's economic policies while continuing to provide credit and having to refuse more credit, thus putting existing loans at risk.

The governments of the Atlantic Alliance, however, appear to face no dilemma and give the impression that they envisage only one alternative: Titoism continues without Tito or the Soviet Union takes all. If that is really what they think, they should prepare for the worst, for Yugoslavia is likely to undergo significant changes in both domestic and foreign policies. If the economic crisis has begun to receive coverage in the Western media, the country's even more serious crisis of political legitimacy has had little attention.[18] The real alternative is not one between the status quo and a return to the Soviet fold. It is a question of knowing which way the collective leadership will turn to try to make up for the authority that has departed with Tito and to what extent its members are likely to agree among themselves. In spite of growing popular disenchantment with the system, and of the readiness of some sections of the leadership to discuss what is wrong with the current situation, Tito's successors have been unable to agree on even the outlines of reform proposals. If the West does not have a clear idea of how it would like Yugoslavia to evolve, the mirage of the status quo will quickly fade away because those who are prepared to defend the present order are also those who fear change even if they sympathize with the aspirations of others who want change. They are also the ones who would be the first to run for help to the USSR if their order started to crumble under public pressure. The Soviet Union, which neither needs to create problems nor to uphold stability, knows how to exploit in its own interest the possibility of change in Yugoslavia.

Yugoslavia's leadership might turn even more toward the army, with greater emphasis on an orthodox pattern of domestic control. Initially, such a trend could be adopted to maintain independence from the Soviet Union, but it would have to strengthen the restrictions on the dynamic forces in society that are pulling toward more independence from the Center and generally pulling toward Europe. Could the leadership, even if it were united, hold out against both the Soviet Union and increased domestic dissension? The likelihood is that the people in power would then let themselves be carried along by a drift that would bring the country back to the side of the USSR, or would even have to turn to the Soviets for help.

On the other hand, the leadership might try to obtain the confidence of a broader consensus for a more open style of government, which could pursue down-to-earth policies in an atmosphere of responsible internal détente. Tito is no longer there to prevent such a trend, put on the brakes, or reverse it.

---

[18]See George Schöpflin, "The Yugoslav Crisis," *Soviet Analyst* 12, no. 2 (26 January 1983): 1–4.

At the very least, it would not worsen existing problems and would not upset the established principles of the union of differentiated nationalities, of Socialist self-management, or of nonalignment with existing military alliances.

The country's limited freedom of choice to evolve this way or that depends on the attitude of the international community. It would be helped by an accepted and acknowledged neutrality. It is only in conditions of relative détente that a country like Yugoslavia can keep a certain margin of independence and a limited amount of choice. The interests of East and West converge, for the time being, to the extent that neither would gain from its collapse in the near future. Yugoslavia, however, would have to renounce what remains of its self-proclaimed and self-deluding role as mentor to the world and concentrate on unraveling its own accumulated difficulties. Without Tito, it will not be difficult to acknowledge the limited nature of leadership in the present world, and it should be easier to acknowledge the limited nature of Yugoslavia's development. But it will need more self-assurance to stop proclaiming socialism as a final solution and to accept it more humbly as a frame of mind for looking at societies from the point of view of the disadvantaged. It would need even more self-assurance to renounce the present sources of legitimacy—the authority of the late Marshal Tito and the partisan struggle in the Second World War—and to find new forms of legitimacy, based on popular support, on the regime's own effectiveness and on its ability to evolve. Those in authority over the past forty years not unnaturally fail to understand that policy conflicts are a normal facet of democracy. They attribute instability to subversion by the joint endeavors of outside and underground evil forces, whereas it is they who have been increasingly responsible for an instability of policy caused by arguments between sections of the establishment.

Yugoslavia's limited freedom of choice would be helped by a more clearly businesslike Western attitude. Constant praise for the Titoist position only causes contempt and an impression of insincerity. It makes Tito's successors, and the population at large, doubt that the West can be relied upon and may encourage them to seek and accept a reassurance from Moscow which is seen, on the whole, as having come to accept Yugoslavia as it is. For a change, the West could think of encouraging those who are looking for a more pluralistic, if less anti-Soviet, regime. Western policy has been inadequate insofar as it has not fully taken into account the need to maximize the prospects for positive change and thus failed to take advantage of the opportunity to cut the grass under Soviet feet.

In spite of blanket references by all U.S. presidents to their interest in Yugoslavia's independence and integrity, the recent American approach has been altogether incoherent. Jimmy Carter's electoral statement of October 1976—that he would not go to war for Yugoslavia even if the Soviet Union sent in troops because this would not threaten American security—was not subsequently disavowed. Coming after the pronouncements of Helmut Sonnenfeldt in December 1975 (regardless of intentions and later emendations), it hardly implied willingness on the part of Washington to act on behalf of

Yugoslavia. President Ronald Reagan's policies have tended to confuse antagonists and alienate allies, but they do not seem to imply the danger of direct aggression in Europe and, to that extent, almost leave Yugoslavia aside. His speech to British members of Parliament in June 1982 and subsequent reports of a "Crusade for Democracy" program are still too embryonic to evaluate. Do they indicate a new realization of the inherent instability of dictatorships (or the danger of their apparent stability), or are they but a gimmick? To what extent might the "crusade" apply to Yugoslavia? As for the European partners of the Atlantic Alliance, they expressed concern at the time of Tito's death, while the European Economic Community rushed to bring to a rapid conclusion its long-protracted trade negotiations with Yugoslavia. Do the Europeans still consider that they have "vital" interests beyond their immediate present borders? If they do, what sacrifices are they ready to make in order to sustain such interests and to what extent does Yugoslavia figure among them?

Yugoslavia's limited freedom of choice would be helped if the Soviet Union knew unambiguously that the Western allies, individually and collectively, regard that country's military neutrality and its political independence as an essential component of the established balance of power. While they would understandably not want to limit their options in advance, committing themselves to the defense of Yugoslavia as it stands at present, it would not be in their interest that Moscow should be more or less free to resort to arms. Yugoslavia needs some form of protection from military intervention if it is to open out rather than close in. A major factor affecting Soviet policy decisions on Yugoslavia is thus likely to be the extent of Western skill in signaling, but the West would first have to decide how much it valued Yugoslavia's ability to remain independent, neutral, and to adapt to changing circumstances.

Yugoslavia's limited freedom of choice might be helped if it were decided that it was also worth some economic sacrifices at a time when the West feels sorry for itself. This need not mean pouring in more cash to keep afloat the blend of extravagance, arrogance, and inefficiency, nor would it mean using trade as a negative political instrument; it could be a "give-and-take." In a businesslike atmosphere, no one could object to advice being given to the Yugoslav leaders on how to approach some of their problems. If everything were made clear, there would be no need to worry so much about a bit of destabilization that would encourage those who want to open out and give the Yugoslavs a chance to grope toward some broadly based accepted stability.

CONCLUSION: THE BALKAN ALTERNATIVE

Dogmatically rather than pragmatically nonaligned, Yugoslavia is dangerously isolated. Would it not suit its interests better to moor somewhere, and to moor where it happens to be located geographically? There are, in historical terms, several aspects of Europe other than the one that has taken shape around Western Europe. One of these is Balkan Europe whose organic realization is, in the foreseeable future, obviously impossible. Nevertheless, Yugoslavia should not find it difficult, as an absolute neutral, to hitch on to the Balkans, a region that cuts across military alliances, most of whose leaders in recent years have expressed the wish to turn the peninsula into an area of peace and cooperation and where Yugoslavia could provide positive services by a really neutral existence.

Yugoslavia's three Communist Balkan neighbors all expressed concern about the international situation at the time of Tito's death. Relations with Romania had been developing well, though never to the point of actual military cooperation. If they have reached a point beyond which it seems they will not mature, this is because of the uncertainty each feels about the other's future, perhaps even the fear each has that the other may be selling out to the USSR.

Bulgaria holds a key position in the Balkans because it can affect Yugoslavia through Macedonia, and because it borders on two members of NATO. Relations between Sofia and Belgrade tend to reflect the state of Soviet-Yugoslav relations, even though there is increasing evidence that the current Bulgarian leadership and its policies are more original than the simplistic view of "the USSR's best satellite" would have it. In the autumn of 1980 there were exchanges of visits between both Romania and Yugoslavia and Bulgaria and Yugoslavia.

Even though relations between Belgrade and Tirana are tense as a result of the latest explosion in Kosovo, there are important limits to the war of words waged between the two in 1981. In its open pronouncements, at any rate, the Albanian government went no further than to express support for the establishment of a full republican homeland, within the Yugoslav federation, for Yugoslavia's ethnic Albanians. As for the Yugoslav government, it never even raised the possibility of retaliating by blocking Albania's foreign trade which has opened up through Yugoslav territory. Both sides share two characteristics: they are insecure, and they are independent of the Soviet Union.

From the southern NATO side, Greece has established links with its Communist neighbors, a policy actually initiated by the Colonels. Worried about future developments in the peninsula, particularly in Yugoslavia, the Karamanlis government was anxious to set up regional ties in order to help stabilize the situation. In November 1980, President Karamanlis visited Belgrade. Relations improved further with the advent of a Socialist government in Athens. In April 1982 the Yugoslav foreign secretary traveled to Athens, and in May Prime Minister Andreas Papandreou was received in Belgrade with exceptional warmth. The two governments pledged to keep on strengthening mutual links and to broaden Balkan regional ideas. Papandreou's government was intent on expanding Greece's active interest in Balkan cooperation and promoting the idea, now revived by the Greek Socialist leader, of a Balkan nuclear-free zone. This received strong support from Bulgarian President Todor Zhivkov and was welcomed in all Balkan capitals except Tirana and—to the extent that Turkey is in part a Balkan country—Ankara. The Turks also have had a round of talks and visits with Balkan Communist governments, including Yugoslavia.

The nuclear-free zone is essentially a Greco-Bulgarian project, to which Yugoslavs and Romanians have responded cautiously. Yugoslavia has reservations about a Balkan summit devoted to it and has argued rhetorically that such a project did not stand a chance unless it was taken in the context of European security, in general, and of Mediterranean security, in particular, and, more realistically, as long as there remained open bilateral issues such as its own dispute with Bulgaria over Macedonia.

Yugoslavia and Greece have maintained close military contacts ever since the restoration of a parliamentary government in Athens, but their nature and frequency mainly have been kept secret, so determined has Belgrade been to remain, and be seen to remain, nonaligned. The Yugoslav Communists' only reservation regarding the Greek Socialists' electoral victory was PASOK's original anti-NATO platform. Far from wishing Greece to move away from the alliance, the Yugoslav government, as of late, has been eager to see it again firmly in the organization. Greece in NATO discourages the USSR from taking risks in the Balkans. Along with Italy, it is also seen as a lifeline for material support in case of need. Papandreou's visit to Belgrade thus had another, more discreet, side to it: to reassure Tito's successors that his government would not pull Greece out of NATO.

The Yugoslavs are as interested as the Atlantic allies in restraining the Soviets. What Sonnenfeldt was trying to point out, more for the sake of the Yugoslavs, was that in view of their international attitude Moscow was bound to exercise more influence on Belgrade than Washington. The Yugoslav government could come to realize that, in a fluid and unstable situation, systematic anti-Americanism is harmful to Yugoslavia, and that the European Community should occasionally be taken more seriously as an advocate of pan-European interests, whose links with Yugoslavia can exercise an important function to counteract its isolation. At the same time, the partners of the Atlantic Alliance could come to realize the need to save the Yugoslav regime from its own self while the situation is still fluid. If the Soviet Union is not to take all when the changes set, it is high time to go beyond the simplistic view that the West's only interest in Yugoslavia is to prevent it from sliding back into the Soviet bloc, and that such an end can be furthered by toeing the Yugoslav line and by open-ended financial support.

For a start, Western representatives should stop encouraging a sclerotic, static, worn-out status quo and try to identify the various tendencies that are appearing in the leadership. Would the Atlantic Alliance, at the very least, not want to help the survival of the present system in Yugoslavia rendered more sound in its economics, more acceptable politically to its subjects, more adaptable to change; in short, more viable in a long-term perspective that looked at the continuity of the national entity rather than at the specifics of the regime? There was one Western head of government who did at least state publicly in Belgrade: "Yugoslavia existed before Tito, and Yugoslavia will exist after Tito."[19]

[19]French Prime Minister Raymond Barre, 14 November 1980.

Greece and NATO

S. VICTOR PAPACOSMA

"We believe that with the support of the people we can attain national pride and a Greece which will exercise a national, independent and multi-dimensional foreign policy. We have but one restriction: the national interest."[1] With these remarks before the newly elected parliament on 22 November 1981, Prime Minister Andreas Papandreou summarized the foreign policy objectives of his young government. Implicit in these comments are a critical assessment of past policy and the determination to forge a new-style diplomatic path. Both attitudes are not, however, original for the Greek scene. It thus remains to be seen whether Papandreou will succeed where others have faltered. Greece's relationship with NATO is obviously one of the foremost components in any foreign policy initiatives.

HISTORICAL PATTERNS

Emerging from its struggle for independence against the Ottoman Empire during the 1820s, Greece found itself strapped into a dependency role in foreign affairs. The great powers considered the Greek-inhabited area of the southern Balkans a valuable piece of real estate, primarily for its strategic location and not necessarily for its human and natural resources. Seeking to maintain the regional balance as closely as possible, the great powers carved out a diminutive Greece, which included barely one-quarter of all Greeks in the eastern Mediterranean. Presiding over a small and desperately poor nation, Greek leaders nonetheless focused their efforts on the liberation of their countrymen still under Ottoman overlordship. But rivalries among Greece's guaranteeing powers—Britain, France, and Russia—and balance of power considerations prevented Greece from advancing its irredentist campaign against the Ottoman Empire. Greece's attempts to assert its own policy, independent of great-power approval, customarily met firm opposition.[2] Moreover, Greece

---

[1] *Greek Government Programme—Presented by the Prime Minister Andreas G. Papandreou* (Athens: General Secretariat for Press and Information, 1981), pp. 5–6.

[2] For example, in order to restrain Greece from siding with Russia during the Crimean War, British and French warships occupied the Piraeus harbor from 1854 to 1857. Two decades later diplomatic pressure from London and Paris prevented Greece from entering the Russo-Turkish War of 1877–78. After Bulgaria's annexation of eastern Rumelia in 1885, the Greeks mobilized, only to find their coasts once again blockaded by the British and French in May 1886.

generally failed in its attempts to exploit great-power rivalries, a tactic that the Ottoman Empire, larger in size and bordering on tsarist Russia, pursued more effectively. With no regional allies and confronted by pressures from the British and French, Greek governments periodically sought rewards for "good behavior." The resulting territorial acquisitions were too few and far between for most Greeks to endorse heartily such a policy. In no position to challenge the powers dominant in the eastern Mediterranean, Greece found itself seeking cooperation and aid from the very states that generally held it in check.

Greek aspirations moved closer toward fulfillment with significant territorial gains during the Balkan wars of 1912–13. Eleftherios Venizelos, the architect of Greece's victories, saw the next opportunity for acquiring more territory emerging during World War I. With minor variations he followed a familiar policy of maintaining a good behavior position vis-à-vis Greece's powerful patrons. While King Constantine I spoke out for continued neutrality, a policy beneficial to Germany's limited capabilities in the eastern Mediterranean, whose waters were dominated by the British and French, Venizelos pressed for Greece's entry into the war on the side of the Entente. Greece split into two political camps but finally entered into the war when Anglo-French pressure forced Constantine to leave Greece in June 1917. For his loyalty to the victorious coalition, Venizelos received promises of considerable territorial gains at the Paris Peace Conference. But the political divisions within Greece, and subsequently between Greece's patrons, Britain and France, revealed the fragility of Venizelos's tactics. Greece gambled and lost, the Turks defeating its forces in Asia Minor during 1922.

Greece and Turkey confronted difficult tasks of national recovery in the years immediately following, and their relations remained strained until the end of the decade. Then in a dramatic turnaround from past patterns, Venizelos and Kemal Ataturk signed treaties of friendship and commerce in October 1930. Both leaders made important concessions, renounced any expansionist designs against each other, and boldly acknowledged the benefits of working together in order to secure joint interests. A spirit of unprecedented cooperation characterized relations during the 1930s, as Greece and Turkey coordinated policies, first in the Balkan Conference movement and later with the Balkan Entente, which also included Yugoslavia and Romania. A serious attempt was made to develop regional collective security pacts and to avoid entangling agreements with the bigger powers. Such a policy proved effective only as long as these powers did not actively infiltrate Balkan affairs. Consequently, regional commitments wilted in the face of Axis advances toward southeastern Europe and particularly after war broke out in September 1939.

Events and policies associated with the Axis occupation, both inside and outside of Greece, thrust the country into the further considerations of the powers. The progressive breakdown of the anti-Axis coalition in the last months of World War II, the dominating presence of Soviet forces in Eastern Europe, and the growing strength of native Communist movements in the region found expression in Greece. Britain's deliberate objective of keeping secure Greece's traditional association with Western Europe played an important part in the Moscow Conference of October 1944 when Churchill and Stalin concluded their "percentages agreement."

Shortly after its liberation in October, Greece became embroiled in fratricidal conflict with leftist-led resistance forces battling Greek government and British troops. Fighting ended in the midwinter of 1945, but during the following year Greek Communists, with the support of neighboring Balkan regimes, began another campaign. By early 1947 London informed Washington that strained resources could not sustain its support of Greece during this troubled period. The U.S. government responded with the Truman Doctrine on 12 March 1947, thereby committing itself to come to the aid of Greece and Turkey against Communist pressures. The Cold War had commenced, and for Greece, an important focal point in this struggle, it meant that the United States undertook the role of patron formerly assumed by Britain and less frequently by France. Coincident with extensive military and economic aid, which contributed to the crushing of the rebellion in 1949, came increased American involvement in Greek affairs.

ENTRY INTO NATO

For the moment Greece's territorial integrity and links with the West had been protected. With their country greatly weakened by nearly a decade of wartime conditions and suffering, Greek leaders next looked to membership in NATO as a more permanent guarantee of its national security against the Communist threat from the North. Concurrently, Greece and Turkey had become outposts against Soviet expansion in the eastern Mediterranean. This joint mission against the common enemy seemed to be drawing the two nations even more closely together than had been the case in the 1930s.

First raised formally in April 1950 with their applications, the proposal for the inclusion of Greece and Turkey into the Atlantic Alliance met initially with opposition and held the firm support only of Italy. Denmark, Norway, and the Benelux states expressed apprehension about assuming responsibility for the defense of a region quite distant geographically and culturally from Western Europe and their own acknowledged strategic interests. Some member-states feared that additional participants would reduce their share of vital American armaments. Britain preferred instead the establishment of a separate Middle East Command. Greece and Turkey, it was argued, were linked strategically with the Middle East, and such an organization would take into consideration Commonwealth contributions and the Arab states, particularly Egypt. London was also concerned about the placing of an American commander over British Mediterranean forces.

At the September 1950 meeting of the North Atlantic Council in New York, it was decided that Greece and Turkey would be accorded associate status and could participate in defense planning relative to the Mediterranean. Opposition to full membership decreased in the following months, largely in reaction to the Korean War and unsettling developments in the Middle East. Moreover, the United States increasingly backed their joint entry in order to secure the southern flank of the SHAPE command and to establish American air bases in Turkey. In May 1951 Washington formally proposed that full membership should be granted to Greece and Turkey. The worsening Middle East situation, and particularly the Iranian crisis, convinced the British in July to add their backing. The Atlantic Council unanimously recommended, although with reluctant approval on the part of Denmark and Norway, the inclusion of Greece and Turkey into NATO at its September 1951 session in Ottawa. Formal entry came at the Lisbon meeting in February 1952. The southern flank experienced further buttressing when negotiations, initiated during 1952, evolved into a treaty of friendship and cooperation with Greece, Turkey, and Yugoslavia in February 1953. A formal treaty of alliance followed in August 1954.[3]

TROUBLED PARTNERSHIP

As part of a broad alliance system to protect its national security, it was expected that Greece could concentrate on internal reconstruction. This would not be the case, however, as old patterns surfaced to torment new relationships. America's already heavy presence increased with the 1953 agreement which provided for the establishment of U.S. bases on Greek soil. The Right readily accepted this development, and the liberal center expressed some partial disagreement, while the Left expectedly offered strong opposition. Once again a weak Greece was linked with the region's dominant powers in a client-patron relationship mindful of earlier decades.

Rather quickly events on British-controlled Cyprus, with its movement for *enosis*, or union, with Greece during the mid-1950s, revealed for many Greeks the unfortunate dimensions of their dependency status. The Cyprus question aggravated relations between Greece and Turkey, revived hostile attitudes, and became an international issue whose impact would be felt

[3]D. G. Kousoulas, *The Price of Freedom: Greece in World Affairs, 1939–1953* (Syracuse: Syracuse University Press, 1953), pp. 184–87; D. J. K., "Greece, Turkey, and Nato," *The World Today* 8, no. 4 (April 1952): 162–67; Hamilton Fish Armstrong, "Eisenhower's Right Flank," *Foreign Affairs* 29, no. 4 (July 1951): 651–63; Joe R. Wilkinson, "Denmark and NATO: The Problem of a Small State in a Collective Security System," *International Organization* 10, no. 3 (August 1956): 395–96; George F. Kennan, *The Cloud of Danger* (Boston: Little, Brown & Company, 1977), pp. 114–15. Details on the negotiations surrounding the Balkan alliance are to be found in John Iatrides, *Balkan Triangle: Birth and Decline of an Alliance Across Ideological Boundaries* (The Hague: Mouton, 1968).

beyond the eastern Mediterranean. This conflict glaringly revealed that, whereas the United States and most of its NATO allies might share a strategic world view and strong commitment to collective security during this Cold War era, the two members of the southeastern flank too easily shifted attention to divisive regional issues.

Conquered by Turkey in 1571, Cyprus came under British administration in 1878. The Greek-Cypriot campaign for *enosis* with Greece, sporadically evident prior to World War II, did not intensify and gain momentum until the 1950s. Greek claims centered on the generally accepted fact that the overwhelming majority of Greek Cypriots, comprising nearly 80 percent of the island's inhabitants, demanded *enosis*. Ankara backed the Turkish Cypriots, about 19 percent of the population, in their desire not to be absorbed by Greece. Widespread Greek-Cypriot terrorism, directed against British and Turkish interests in Cyprus in 1955, provoked Turkish reprisals against the Greek community in Istanbul, causing most of its members to leave in the months and years that followed. In the aftermath of these unfortunate developments, Greece temporarily withdrew its officers from NATO's regional headquarters in Izmir, and the recently concluded Balkan Pact with Turkey and Yugoslavia became a dead letter.

Britain's role in exploiting differences between Greeks and Turks on Cyprus in order to maintain its control heightened resentment among Greeks. For the historically minded in Greece, it evoked memories of Britain's nineteenth-century policy to sustain the Ottoman Empire against Athens's efforts to liberate Greek-inhabited lands. Even conservatives spoke of possibly withdrawing from NATO rather than seeing Greece's interests sacrificed by supposed allies.

Following the failure of the United Nations to provide a political settlement, Lord Ismay offered to serve as mediator for all three parties involved: the British who had secured the island from Turkey in 1878, the Greeks who supported the joining of the island to Greece, and the Turks who were concerned over their own proximity to a possibly Greek-controlled Cyprus and over the fate of the Moslem minority of Turkish descent. The three powers met in Paris in 1957 and failed to find a consensus. Ismay's successor, Paul-Henri Spaak, tried again the following year with a "Spaak Plan" for a federated arrangement with separate Greek and Turkish legislative bodies that would advise the British governor for a seven-year period. Although this plan also foundered, NATO could at least take credit for an initiative that led to a temporary resumption of bilateral Greek and Turkish conversations, eventually leading to uneasy compromise in the Zurich agreement of 1959. Cyprus became independent in 1960. The Greeks did not win *enosis*, and the Turks did not win partition but received special safeguards to protect their cultural and political interests. The only overt American role in this period was President Dwight D. Eisenhower's public support for Ismay's efforts in 1957.

NATO appeared to profit from the arrangements. Greece, Turkey, and the United Kingdom were all granted special rights in Cyprus—the British with their two bases and the Greeks with 950 troops and the Turks with 650.

The common concern for the protection of the eastern Mediterranean seemed to prevail over national passions. Still other disintegrative issues surrounding Greece's participation in NATO included concern over high domestic expenditures for the Greek armed forces and the installation of nuclear weapons on Greek soil. By the end of the 1950s, the liberal center had come to question more openly Greece's exclusively one-sided foreign policy alignment, while the Left advocated a neutralist alternative. With the establishment of Cyprus's independence, it was expected by hopeful observers that Greece's links with NATO would stabilize once again.[4]

Cyprus, however, did not shrink into the diplomatic background. Instead, this small island-state persisted as a multifaceted international problem. Disputes between Greek and Turkish Cypriots would poison relations between the two countries. Concurrently, the geopolitical scene changed as the United States and NATO found Cyprus more and more important. The loss of other bases in the Mediterranean placed new emphasis upon this island at the very time Britain's own power was suffering continuing attrition. In this context, the Americans assumed an even larger role with Athens and Ankara separately seeking diplomatic backing and moral support from the United States and NATO in their disputes.

In late 1963 Cyprus's president, Archbishop Makarios, proposed thirteen constitutional amendments to streamline some of the nation's unwieldy institutions and procedures. Some of the proposed changes sought to curb the veto power of the Turkish minority in legislation and to whittle down the representation of the Turks in the governmental services to a level more representative of their share of the total Cypriot population. Fighting between the two communities broke out, and Ankara threatened to invade the island. Makarios rejected an Anglo-American initiative for a NATO peace-keeping force and mediation, anticipating that Greek-Cypriot interests would be sacrificed for those of Turkey and the alliance. As an alternative, a UN peace-keeping force arrived in March 1964, but the volatile atmosphere persisted. President Lyndon B. Johnson sent a tersely worded note to Turkish President Ismet Inönü in June 1964, warning him of dire consequences if Turkey chose to invade the island.[5] Inönü backed off, but the Turks would not easily forget the reprimand. Turkey subsequently assumed a more critical position toward American policies and modified its formerly staunch pro-NATO stance. Concurrently, Athens and Nicosia aired their own set of grievances in response to NATO pressures, specifically to the "Acheson Plan." The former secretary of state, acting on behalf of the American president, had devised a plan for union between Greece and Cyprus in return for the creation of a Turkish

[4]Information on the development of public opinion and political positions is located in Theodore A. Couloumbis, *Greek Political Reaction to American and NATO Influences* (New Haven: Yale University Press, 1966).

[5]The full text of President Johnson's letter of 5 June 1965 was published in January 1966 in the Turkish newspaper *Hurriyet*, 14 January 1966. See A. H. Ulman and R. H. Dekmejian, "Changing Patterns in Turkish Foreign Policy, 1959–1967," *Orbis* 11, no. 3 (Fall 1967): 778.

military base and Turkish-Cypriot cantons on Cyprus and the cession of the small island of Castellorizo to Turkey.

Greek internal politics became increasingly turbulent after the spring of 1965, only to be quashed with the imposition of a harsh military dictatorship on 21 April 1967. The conspirators implemented a NATO contingency plan, code named "Prometheus," designed for the event of serious internal disorders. While public opinion and some governments in Western Europe regularly criticized the Greek junta, it never experienced serious pressure from NATO. The Colonels carefully fulfilled Greece's obligations under the NATO alliance. In turn, American and NATO officials visited Greece on a regular basis. Despite the enactment on 16 May 1967 of an embargo on heavy arms, the United States continued to be Greece's major source of military equipment. During a period of constant turmoil in the Middle East and of an augmented Soviet presence in the Mediterranean, the stability offered by the dictatorship appeared to serve American interests. The signing of a home-port agreement in 1972 for the stationing of U.S. Navy carrier forces near Athens provided another example to the dictatorship's critics of America benefiting from the authoritarian regime's existence.

The junta faced its first international crisis in November 1967 when intercommunal fighting once more erupted on Cyprus. Again, American diplomatic pressure averted pitched battle between the two NATO allies. By this date, too, old sentiments had undergone an evolution. The desire for *enosis* had considerably diminished among Greek Cypriots who supported Makarios's policy of protecting flourishing Greek interests within the context of an independent Cyprus. Fervor for union on the mainland also waned. Conditions would remain relatively quiet on the island for several years, although no real progress was made in intercommunal talks. In turn, relations between Makarios and the Greek dictatorship worsened.

In November 1973, Brigadier Demetrios Ioannides engineered a putsch to recast the Greek junta into a more authoritarian mold. The Ioannides regime from Athens backed terrorist initiatives against Makarios who was accused of betraying *enosis*. Tension increased between Athens and Nicosia. Finally, under the superficial camouflage of an internal coup d'état by the Greek-Cypriot National Guard—dominated by Athens—against its president, the Ioannides junta forcefully brought about Makarios's ouster on 15 July 1974.

The extreme nationalism and anticommunism of Brigadier Ioannides's coterie motivated these minority interests to score what they thought would be a diplomatic triumph, thereby diverting attention from internal problems and also ridding Greek politics of the independent "red archbishop." But the historical feelings for *enosis* no longer galvanized the majority of Greeks on the mainland and on embattled Cyprus. Consolidating its hold on the island, the newly installed outlaw regime of Nikos Sampson found itself recognized only by Athens. Even the Turks, firmly established foes of Makarios, demanded his return to power.

The pace of events quickened. Concerned for its people on a Cyprus controlled by pro-*enosis* elements, Turkey distorted its rights as a guarantor state in the 1959 Zurich-London agreements to suit its specific interests and landed forces on the island in the dawn hours of 20 July 1974. Athens responded with a call for general mobilization, which rapidly became a fiasco. Totally discredited and diplomatically isolated, the Ioannides junta, under pressure from fellow officers, quickly exited on 23 July, hoping to avert national catastrophe through the creation of a civilian government under Constantine Karamanlis.

WITHDRAWAL FROM NATO's MILITARY COMMAND

These swift changes in Greece's government could not clear the domestic scene of all outstanding problems. Despite the initial exhilaration of the Greeks over the dictatorship's downfall, the hangover of Cyprus remained with its continued bloodshed and thousands of refugees. The Turks violated cease-fires, methodically increased their troops on the island to 40,000, and then on 14 August sidestepped the Geneva peace talks with a broad offensive, taking over nearly 40 percent of the island within two days.

Barely four hours after Turkey launched its invasion, Karamanlis announced the withdrawal of Greece's armed forces from the NATO military command. The prime minister later divulged that only two choices lay before him: to declare war against Turkey or to limit links with NATO; he chose the lesser of evils. The government had to yield to public opinion, inflamed by NATO's indifference to Turkey's actions on Cyprus and by charges that NATO acted to limit Greece's response to the Turkish invasion. For years, Karamanlis argued, NATO's leadership had tried to take Cyprus out of UN debates and to coordinate an allied-sponsored solution. With the opportunity to act positively and to avoid tragedy, the Atlantic Alliance instead sat back and allowed Turkey to advance its interests on the troubled island. Karamanlis believed that Greece's military withdrawal from NATO would draw greater world attention to Cyprus's plight and induce NATO allies to work for a just solution. He also needed to have full control over Greece's armed forces during a time of threatening war.[6]

Anti-junta Greeks, who had earlier condemned the United States and NATO for their toleration of and extension of military aid to the dictatorship, now found their ranks swelling by those who perceived Washington as a prime conspirator behind the July coup on Cyprus and as a supporter of Turkish actions. The blundering of American policy toward Cyprus during these confused days immediately before and after the resignation of President Richard M.

---

[6]C. M. Woodhouse, *Karamanlis: The Restorer of Greek Democracy* (Oxford: Clarendon Press, 1982), pp. 217–19; Demetris Bitsios, *Pera apo ta synora, 1974–1977* [Beyond the frontier, 1974–1977], (Athens: Hestia, 1983), pp. 203–05; Theodore Couloumbis, *The United States, Greece, and Turkey: The Troubled Triangle* (New York: Praeger, 1983), pp. 101–02, n. 52.

Nixon has been recorded elsewhere and, most sharply, by Laurence Stern in *The Wrong Horse* (1977). Even without the midsummer events of 1974, the United States and its NATO allies could have anticipated an outburst of Greek sentiment against them on the fall of the dictatorship. A study mission report of February 1974 for the U.S. House of Representatives Committee on Foreign Affairs bore the ominous title: "Controlling the Damage: U.S. Policy Options for Greece." It concluded that the United States had pursued a faulty policy since 1967: "Damage has already occurred to American interests in Greece and more will occur before the present situation ends." The assessment commented prophetically that "it would be highly unrealistic to think that the United States will ever be able to return to the comfortable patron-client state relationship it once enjoyed with Greece."[7]

All the protagonists in the unfortunate scenario were supposed allies, and once again Greece appeared to suffer for its undue dependency on stronger "friends" who willingly seemed to sacrifice Greek national interests. Whereas before the Colonels established their dictatorship, the Left had consistently opposed Greece's close ties with the United States and NATO, and the Center had become increasingly critical of them, since 1974 the Right also has joined the ranks of those questioning the value of these affiliations.

THE AEGEAN

Hostile relations between Greece and Turkey over Cyprus, which had almost resulted in full-scale war, also furnished the occasion for Turkey to challenge Greek rights on still another front: the Aegean Sea. These new issues further complicated the Greco-Turkish imbroglio and, as with the Cyprus conflict, affected broader interests beyond those of the two nations. In a period of regional détente between neighboring NATO and Warsaw Pact states, the Greeks and Turks concentrated their strategic concerns and militancy on each other. NATO solidarity inevitably suffered.

The Aegean has attracted many conquerors and invaders over the centuries. Even before the recording of history, Greeks formed the majority of the inhabitants on its islands, and, despite the succession of foreign occupiers, they maintained this status into the present century. After they won their independence from the Ottoman Turks during the 1820s, the Greeks sought to wrest the eastern Aegean islands and other Greek-inhabited territories from Turkish control. The opportunity came during the Balkan wars when Greece's navy quickly dominated the Aegean and troops were landed on most of the eastern Aegean islands. The Turks worked to reestablish their influence in the Aegean immediately after their defeat in the Balkan wars, but the outbreak

[7]U.S., Congress, House, *Controlling the Damage: U.S. Policy Options for Greece* (Report of a Study Mission to Greece, 18–21 January 1974), 93d Cong., 2d sess., 1974, pp. 1, 12.

of World War I diverted their strategic concerns. With the exception of World War II, the years since World War I had not provided the appropriate diplomatic or military conditions for a political offensive on Aegean matters. Turkish leaders, always concerned about these foreign-controlled islands lying so closely off their coast, intensified such a campaign during the summer of 1974.

The Sea

The first of several dimensions to the Aegean question, that of the Aegean continental shelf, attracted scant attention at first. In 1970 the Greek military government granted a license to a foreign consortium of prospecting companies to search for oil in the northern Aegean. Except for a substantial deposit of natural gas off the west coast of Thasos in December 1972, exploratory drilling produced few results until January 1974. Preliminary reports projected that this find could provide close to 10 percent of Greece's oil requirements. In a period of rapidly rising energy prices, even this small field seemed vital for a nation totally dependent on oil imports and offered hope for new discoveries.

Meanwhile, on 1 November 1973 Turkish authorities published in the *Government Gazette* the award of exploration and exploitation rights in twenty-seven maritime areas of the Aegean to the Turkish State Petroleum Company. The map accompanying the announcement unilaterally delimited the seabed between Greece and Turkey with a median line equidistant from the Turkish and Greek mainland coasts. A number of exploration areas lay just beyond the territorial waters of the populated Greek islands of Samothrace, Lemnos, Aghios Evstratios, Mytilene, Chios, Psara, and Antipsara.

The Greek Ministry of Foreign Affairs charged that Turkish policy did not respect the sovereign rights of Greece regarding the continental shelf of islands, secured by the 1958 Geneva Convention on the Continental Shelf and ratified by Greece in 1972. Ankara argued, in turn, that Turkey had not ratified the convention, adding that the Aegean with hundreds of Greek islands and islets, some lying very near the Turkish coast, created a unique situation not adequately considered by the Geneva Convention. In late February 1974 Ankara proposed that a bilateral agreement on the delimitation of the Aegean continental shelf should be negotiated by the two states. Saber rattling and sharp words characterized relations between the two states during April and May.

Greece agreed on 25 May to the proposal for negotiations "in accordance with international law as codified in the Geneva Convention." Although Prime Minister Bülent Ecevit cited this as a "positive development," the Turkish government announced three days later that the *Candarli*, a survey ship, would sail into the Aegean for exploration studies. On 29 May a small armada of thirty-two warships accompanied the *Candarli* into areas in which the Turkish government had granted mineral exploration concessions. Turkey rejected a

Greek protest. One month later Ankara granted more exploration licenses for sites farther west and south, including waters all around the Dodecanese Islands.[8] It was during these mid-July days that the Cyprus crisis erupted, and the continental shelf dispute faded temporarily into the background.

One authority has aptly summed up the Aegean dispute as "nightmarishly complex, even by law-of-the-sea standards."[9] From the onset of the crisis, Greece has assumed a status quo position by relying specifically on the 1958 Geneva Convention. This agreement states that islands, like the mainland, have continental shelves; that is, submerged off-coast lands extending between the outer limits of a nation's territorial waters and a water depth of 200 meters or beyond that to a depth at which the water above the submerged lands permits exploitation of those lands. Since the Aegean is clustered with hundreds of Greek islands, Greece can claim much more of the Aegean continental shelf. In response, the Turks have taken a revisionist stance, denying that the Geneva Convention is declaratory of general international law. Turkey countered that the Greek islands were not entitled to a continental shelf of their own because they rest on the continental shelf of Asia Minor. Ankara thereupon drew a median line through the Aegean seabed between the mainlands of the two countries, which placed the eastern Aegean and Dodecanese Islands within the purported Asia Minor continental shelf. Relying on existing international maritime law and practices, Greece has stressed a legal solution to the problem. By emphasizing what it considers to be the unique nature of the Aegean question, Turkey has sought a political solution through bilateral negotiations.[10]

The heightened tensions of the summer of 1974 rigidified the positions of both governments and, sequentially, new Aegean-related issues (mentioned below) came to the fore. On the heels of Turkey's mid-August advance on Cyprus, Greece began to fortify its islands off the Turkish coast. Fearing a possible Turkish invasion of one or more islands, Greece, with its defensive measures, appeared to violate the Treaty of Lausanne (1923) and the Treaty of Paris (1947) which forbade the militarization of the Aegean and Dodecanese Islands immediately opposite the Turkish coast. The Greeks saw no other recourse than to secure these islands, particularly in light of bellicose statements made by Turkish officials.[11] Greek fears were further compounded with

---

[8]Alexis Phylactopoulos, "Mediterranean Discord: Conflicting Greek-Turkish Claims on the Aegean Seabed," *International Lawyer* 8, no. 3 (July 1974): 431–41; Andrew Wilson, "The Aegean Dispute," *Adelphi Papers*, no. 155 (London: Institute for Strategic Studies, 1978), pp. 5–6.

[9]Barry Buzan, "A Sea of Troubles? Sources of Dispute in the New Ocean Regime," *Adelphi Papers*, no. 143 (London: Institute for Strategic Studies, 1978), p. 29.

[10]Peter Forster, "Greece and Turkey: The Aegean Conflict," *Swiss Review of World Affairs* 25, no. 12 (March 1976): 10–12; Buzan, "Sources of Dispute in the New Ocean Regime," pp. 29–30; Wilson, "The Aegean Dispute," p. 6.

[11]For example, Prime Minister Sadi Irmak on 13 January 1975 stated: "Half the Aegean is ours. Let the whole world know that this is so. . . . We know how to crush the heads of our enemies when the prestige, dignity and interests of the Turkish nation are attacked." Six days

the establishment of the Fourth Army with headquarters in Izmir. With no NATO commitments and with numerous amphibious landing craft, this new command came to be known as the "Army of the Aegean" and was obviously established to cope with any dangerous escalation of tensions with Greece.

The government of Prime Minister Karamanlis maintained a restrained policy despite pressures from opposition parties and provocative snipes from Turkey. On 27 January 1975 Greece proposed that the two countries should jointly submit the continental shelf issue to the International Court of Justice, a move accepted by Prime Minister Sadi Irmak on 16 February. During the late spring they agreed that legal experts should begin meeting to formulate a seabed compromise for joint submission at The Hague. But shifting governments and opposition politics in Turkey precluded consistent policy statements and actions. Turkey's use of creative legal arguments and diplomatic maneuvers, however, reinforced its general quest for a political settlement that could produce important regional gains. Greece persisted in its legal course since existing laws and practices favored its position; it also realized that the two sides would never be able to resolve all their differences by themselves. Moreover, a legal decision by the International Court of Justice would probably elicit less of an outcry from the Greek public.

Little progress occurred. Then in February 1975 Ankara announced that a hydrographic research ship, *Sizmik I*, was being prepared to prospect in the Aegean; exploration licenses covered all Turkish claims to the continental shelf. While the *Sizmik* was being fitted, Greece on five occasions protested its mission. Athens warned that if the ship entered areas claimed by Greece a dangerous situation would result. On 6 August 1976 *Sizmik*, accompanied by a minesweeper, sailed its way into waters west of Mytilene, which Greece considered part of that island's continental shelf. A war of nerves ensued, and Greek leaders later maintained that war almost broke out. Indeed, Papandreou in the political opposition called for the ship's sinking. The *Sizmik* returned to port on 10 August, the same day that Greece asked for an urgent meeting of the UN Security Council and instituted proceedings in the International Court of Justice against Turkey.[12]

In response to arguments by both sides, the Security Council adopted Resolution 395 on 25 August 1976. The council, determined not to deal

later *Cumhuriyet*, the prominent Istanbul newspaper, quoted Foreign Minister Melih Esenbel as stating that "the policy followed by the growing Turkey to safeguard her interests in the Aegean has to be different from the policy of the Turkey of 50 years ago. . . . Turkey is not the Turkey of 1923 after the Lausanne Treaty. It is a growing Turkey." Prime Minister Suleyman Demirel, in an interview with *Paris Match* on 5 July 1975, declared: "Look at the map. Does the Aegean Sea look like a Greek lake? Furthermore, history bears out what geography tells us. Up to recent times, the islands of the Aegean Sea have always belonged to those who possessed Anatolia." Thanos Veremis, *Greek Security Considerations: A Historical Perspective* (Athens: Papazisis, 1980), p. 62; Wilson, "The Aegean Dispute," pp. 7, 39.

[12]Leo Gross, "The Dispute Between Greece and Turkey Concerning the Continental Shelf in the Aegean," *American Journal of International Law* 71, no. 1 (January 1977): 31; Wilson, "The Aegean Dispute," pp. 8–10.

directly with the substance of the conflict, instead provided a procedural framework for a settlement. It called on the two nations "to do everything in their power to reduce present tensions in the area so that the negotiating process may be facilitated." The resolution also invited Greece and Turkey to take into account the contribution that "appropriate judicial means, in particular the International Court of Justice, are qualified to make in the settlement of remaining legal differences which they may identify in connexion with their present dispute." The Security Council resolution, while somewhat vague, did manage temporarily to dampen the dispute. In turn, the International Court of Justice also provided inconclusive results. Ultimately, in December 1978 the court ruled that, since Turkey refused to be a party to the process, it could not take jurisdiction in the *Aegean Continental Shelf* case.[13]

In November 1976 experts from Greece and Turkey agreed at a Berne meeting to establish a code of behavior to govern future negotiations on the continental shelf. They concurred that talks should be sincere, detailed, and strictly confidential in order to reach an agreement based on mutual consent regarding the continental shelf. They also undertook to abstain from any initiative or act that would prejudice negotiations.[14]

Subsequent meetings, however, provided few results. Obstructionist party politics in both nations, notwithstanding, broader issues beyond the continental shelf have complicated the negotiating process. Fundamentally, Athens and Ankara remain concerned about sovereignty questions. Turkish insistence on a demarcation line for the continental shelf midway between the Greek and Turkish mainlands intensifies Greek fears. Athens sees this demarcation line as a direct threat to the sovereignty of the many Greek islands lying east of it. Turkey would have justification for establishing an economic zone with installations and then a security zone in the seas around these islands. The Greeks hold that these islands form a political continuum with the Greek mainland and that with increased regional penetration the Turks would be able to interfere with Greek internal sea and air communications. The islands would then become threatened enclaves whose sovereignty would be menaced. This Greek archipelago, with its 330,000 inhabitants, is politically and economically linked with the mainland. As evidence that such apprehensions are not mere paranoia, the Greeks cite the invasion of Cyprus and the formation of the Army of the Aegean, whose reported numbers approach those of the entire Greek army. From its vantage point, Turkey argues that sovereign rights for Greece in the continental shelf really will make the Aegean a Greek lake.[15]

---

[13]Richard T. Robol, "Limits of Consent: The *Aegean Sea Continental Shelf* case," *Harvard International Law Journal* 18, no. 3 (Summer 1977): 649, 655–58; Gross, "Dispute Between Greece and Turkey," pp. 39–56; Wilson, "The Aegean Dispute," pp. 9–10.

[14]Text of Berne Declaration in Wilson, "The Aegean Dispute," p. 29.

[15]Marian K. Leighton, "The Antagonists: Greece and Turkey," *Conflict Studies*, no. 109 (July 1979): 5; Wilson, "The Aegean Dispute," pp. 13–14, 22–23.

Turkey is also concerned that Greece might extend its territorial waters from six to twelve nautical miles. Since before World War II Greece has adhered to a limit of six miles, which Turkey also maintains in the Aegean, although it has a twelve-mile limit in the Mediterranean and Black seas. Because Greece has more than two thousand islands in the Aegean, about 35 percent of the region is designated as Greek territorial sea and 8.8 percent as Turkish territorial waters. If Greece followed the example of many maritime states and extended its limit to twelve miles, the Greek portion of the Aegean would increase to 63.9 percent and Turkey, responding similarly, would only have 10 percent. High seas would shrink from 56 percent to 26.1 percent, and all ships leaving Turkish Aegean ports for the Mediterranean would have to pass through Greek waters. Turkey believes that the right of innocent passage is an inadequate guarantee and that it is too vulnerable to total enclosure. Ankara therefore threatens that any Greek extension of territorial seas to the twelve-mile limit will be a casus belli. Greece has not made such a statement but will not forsake its right to do so, a right confirmed by the 1982 Convention on the Law of the Sea.[16]

The Air

Sovereignty issues also relate to the dispute over Aegean airspace which flared up during July-August 1974. In 1952 the International Civil Aviation Organization (ICAO) had arranged with Greece and Turkey for the establishment of a Flight Information Region (FIR) demarcation line dividing the Athens and Istanbul FIR along a line roughly equal to the outer edge of the Turkish territorial sea. Greece thus had responsibility for civilian and military air traffic over almost all of the Aegean. To have placed the FIR line farther to the west would have required Greek aircraft to pass through a Turkish control zone on flights to and from Greek islands in the eastern Aegean.

An agreement that worked well for more than two decades, it was first challenged by the Turks immediately following their invasion of Cyprus. On 6 August 1974 Ankara issued "Notice to Airmen" (NOTAM) 714—a notice to the ICAO for transmission to all air users—requiring all aircraft to report their position and flight plans on reaching the median line of the Aegean. This point, considerably west of the long-established FIR line, was justified, according to the Turks, in order to forewarn them of potential military attack from the west. Greece responded quickly on 7 August with NOTAM 1018, denying the validity of the Turkish act as contrary to existing practices. This was followed by Greek NOTAM 1157 on 13 September, declaring Aegean airspace a danger zone. Since Greece no longer accepted responsibility for

---

[16]Duyhu Bazoğlu Sezer, "Turkey's Security Policies," *Adelphi Papers*, no. 164 (London: Institute for Strategic Studies, 1978), p.17; Buzan, "Sources of Dispute in the New Ocean Regime," p. 29; Veremis, *Greek Security Considerations,* p. 60; Wilson, "The Aegean Dispute," pp. 5, 23.

guaranteed traffic information or safety measures, international airlines suspended all direct flights between the two countries. The new reporting line in the Aegean demanded by the Turks coincided with the western edge of Turkish claims to the continental shelf. Athens perceived this as part of a coordinated campaign against its sovereign rights in the region and its islands.[17] Attempts to settle the FIR dispute and to reopen the Aegean for international flights commenced soon after Greece's issuance of NOTAM 1157. Negotiations, which seemed to be making some progress by the summer of 1976, stalled when Turkish representatives presented a new proposal in March 1977 calling for joint control of all international airspace over the Aegean.

Still another element in the airspace controversy concerns the fixing of Greece's territorial airspace at ten nautical miles. A Greek presidential decree in 1931 had extended the zone from six to ten miles. Turkey had respected this policy until early 1975 when it launched a succession of penetrations by its military planes in the zone between six and ten miles. Greece regularly protested these violations.

After the spring of 1977 talks on these issues made little headway, but on 22 February 1980 Ankara announced that it canceled NOTAM 714 as a goodwill gesture, adding several days later that this initiative did not mean that Ankara would relinquish any of its rights over the Aegean. On 23 February Greece withdrew its NOTAM 1157, and Aegean airspace was opened once again to international air navigation. Whatever optimism the Greeks might have nurtured about an increased easing of relations was soon dispelled when Turkey resumed military flights through Greek airspace contested by Ankara.[18]

NATO

All of these problems between Greece and Turkey also have influenced their roles in NATO and their relations with the United States. The Cyprus question had long plagued NATO and the cohesion of its southeastern flank. The events of midsummer 1974 worsened an already bad situation. War had been prevented, but Greece withdrew its military from NATO commands. Athens also terminated the recently acquired home-porting rights of the U.S. Sixth Fleet. Then in February 1975 Congress imposed an arms embargo on Turkey for having violated the terms of the Foreign Assistance Act by using American arms and equipment to invade Cyprus. Ankara retaliated in July

[17]C. Z. Sazanides, "The Greco-Turkish Dispute over the Aegean Airspace," *Hellenic Review of International Affairs* 1, no. 1 (1980): 88–90; U.S., Congress, Senate, Committee on Foreign Relations, *Turkey, Greece, and NATO: The Strained Alliance,* 96th Cong., 2d sess., 1980, pp. 57–58; Wilson, "The Aegean Dispute," p. 6; Sezer, "Turkey's Security Policies," p. 16.

[18]Wilson, "The Aegean Dispute," pp. 11–12; Sazanides, "The Greco-Turkish Dispute over the Aegean Airspace," pp. 90–97; *Turkey, Greece, and NATO*, pp. 58–59.

by suspending all military arrangements with the United States and halting the activities of almost all American bases in Turkey.[19]

In December 1974, Karamanlis stated that Greece would not rejoin the military wing of NATO until the reasons for its withdrawal were rectified: the Turkish occupation of Cypriot territory and the threat to Greek sovereign rights in the Aegean. Yet, within fourteen months of the Turkish invasion in October 1975, the Greek government announced that it would return its armed forces to the NATO fold on the basis of a "special relationship." Athens contemplated a future role similar to that of Norway, whereby Greek military units would remain under national control during peacetime and revert to NATO direction during wartime.

This turnabout in attitude was largely attributable to Greek concern that Turkey could better advance its claims in the Aegean with Greece out of the military command. Operational control now provided one more contentious dimension to the Aegean disputes between the two states. Athens called in general terms for a resumption of operational responsibilities in the Aegean as they had existed before Greece's withdrawal in 1974. Turkey thereupon emphatically informed its NATO colleagues that it could not consider the restoration of the earlier conditions.[20]

Deliberations on Greek reintegration, which began during 1976, moved slowly. Until the departure of Greek forces from the NATO command, there were two NATO headquarters in Izmir—Land Forces South-East Europe and the Sixth Allied Tactical Air Force—each under an American general with Greek and Turkish deputies. The Greek air force commander had operational control of the airspace from the western coast of Greece to the FIR boundary line just off the west coast of Turkey. The Greeks therefore controlled air defense over most of the Aegean. The alliance in December 1977 announced that the two headquarters would be taken over by Turkish generals with U.S. deputies since, with the withdrawal of Greek forces, it was anomalous for American generals to command only Turkish units. The two commands became "Turkish-NATO" headquarters in July 1978. Prior to this changeover, the alliance gave serious attention to the creation of a "Greek-NATO" headquarters in Larissa, where Greek army and air force commands were already located. Regarding naval matters, Greece before 1974 had operational responsibility in the Aegean (COMEDEAST), while Turkey had a similar function for Turkish territorial waters and the Black Sea (COMEDNOREAST).

---

[19]Turkey requested the United States to suspend operations at all major American installations except the Incirlik air base where American aircraft also have NATO missions. President Jimmy Carter, reversing his 1976 campaign position, succeeded in formally ending the embargo in September 1978.

[20]Ronald Meinardus, "Griechenlands gestörte Verhältnis zur NATO," *Europa Archiv* 25 (February 1982): 105–08; *To Vima* (Athens), 16 March, 5 and 21 October 1980; Veremis, *Greek Security Considerations*, p. 75; *Turkey, Greece, and NATO*, p. 59. During the six-year period Greece remained out of NATO's military structure, it was not a member of the NATO Defense Planning Committee and did not assign troops to NATO commanders. Greece was on the NATO Military Committee, the Nuclear Planning Group, the Budget Committee, the High Level Group, and the Special Group. Greece had representatives in all NATO commands except those in Izmir.

Increasingly, NATO circles talked of developing a "task force" concept that would do away with all area commands in the Mediterranean, thereby undercutting Greece's primary role and increasing Turkey's presence in the Aegean.

In his capacity as NATO's Supreme Allied Commander, General Alexander Haig, in May 1978, submitted a plan for Greece's reintegration to General Ioannis Davos, chief of the Greek General Staff. Few precise details of this and subsequent plans have reached the public eye, but basic information has appeared, mainly in the Greek press. The "Haig-Davos Plan" called for the establishment of a new Greek command in Larissa which would take operational control on an interim basis over the entire Aegean up to the traditional FIR line. This plan would have established a command structure for the Aegean similar to the pre-1974 system except that the Greek command would be situated in Larissa, not Izmir, and a U.S. commander would not be located in the same headquarters. Greece agreed to this scheme for reentry, and all NATO states accepted it except for Turkey which cast a veto. Ankara contended that the interim arrangements for operational responsibilities would become permanent with Greek reintegration since Greece could exercise an effective veto over any later modifications. The Turks essentially argued "either new command boundaries or no boundaries" on airspace questions, while also supporting the task force concept for naval strategy.

Haig produced a second proposal on 6 May 1979, which he transmitted to Davos during an airplane meeting in Verona, Italy. This plan, decidedly more favorable to Turkish interests, called for the reduction of Greek national airspace for NATO military purposes from ten to six miles and the establishment of a task force for Aegean naval defense. A complex scheme arranging new boundaries for the command and control of Aegean air defense provided the possibility of temporary Turkish operational control over tactical command in airspace above Greek islands. Such a contingency, even if coordinated by third-country NATO officers, could not be accepted by Greek leaders.[21]

At the end of May 1979, Haig forwarded a modification of the "Verona Plan" to Davos. This arrangement proposed that a NATO regional operations center in Larissa should govern operation control in Aegean airspace outside Greek national airspace. In this manner, Greek sovereignty over the islands would be preserved, while leaving some role for the Turks outside Greek airspace. It was pointed out, however, that "under this formula a Warsaw Pact bomber flying from Bulgaria to Crete would fly in and out of Greek airspace up to eleven times, and hence Greek officials believe that coordination problems would prove unacceptable."[22]

General Bernard Rogers, replacing Haig as Supreme Allied Commander in late June 1979, submitted yet another variation of Haig's Verona Plan to Davos on 8 November. Athens turned down this version, which included a proposal for no air operational limits over the Aegean. Rogers followed with

---

[21] *To Vima*, 16 March, 5 and 21 October 1980; *Turkey, Greece, and NATO*, pp. 59–60.
[22] *Turkey, Greece, and NATO*, p. 60.

a new plan that was handed to the Greeks on 11 February 1980. Among its provisions were many conditions that Athens thought it could not accept. Quickly evaluating it as the worst in the series of plans, Athens officially turned it down on 22 February as "incompatible with Greek sovereign rights."[23]

In June Athens publicly noted that the status of U.S. military bases on Greek soil was directly associated with the outcome of Greek-NATO negotiations. If the latter failed, then the U.S. bases would have to be closed. Ankara, in turn, continued to stress that the delimitation of the command and control boundaries in the Aegean had to precede Greek reintegration. An important shift did occur, however, shortly after the Turkish armed forces executed their coup d'état on 12 September 1980. A Greek newspaper's interview with the new Turkish foreign minister, Ilter Turkmen, disclosed that Turkey would consider Greece's reintegration first and then work on delimiting the operational boundaries in the Aegean. Conditions up to 1974 would not be binding; until the settlement of the problem, ad hoc agreements could be concluded.

Several reasons help to explain Turkey's change in policy. The new military government sought to concentrate on mammoth domestic problems, the primary motive for its takeover. Turkey also shared with other NATO states considerable concern over developments in Iran, the Soviet invasion of Afghanistan, and the ongoing instability in the Middle East. A unified southeastern flank would secure the interests of all NATO partners. Doubtlessly, America exerted some pressure on the new Turkish regime to yield on major points. Moreover, the prospect of a victory by Papandreou, a bitter critic of NATO, in elections scheduled for 1981 might eliminate any prospect of Greece's reintegration.

Intense activity followed the announcement of this important Turkish policy shift. NATO officers met with officials in Greece and Turkey to produce a new compromise document. On his return to Brussels from Ankara on 17 October 1980, General Rogers landed in Athens to meet with General Agamemnon Gratsios, chief of the Greek General Staff, and to relay the latest proposal already accepted by Ankara. Greece endorsed the plan on the 18th, and two days later NATO's Defense Planning Committee unanimously approved the reintegration of Greece into the alliance's military command. The Greek parliament provided the setting for heated debate, but the New Democracy government with its majority successfully survived the vote of confidence resolution on reintegration on 24 October.

[23]Sazanides, "The Greco-Turkish Dispute over the Aegean Airspace," p. 99; *To Vima*, 16 March 1980; C. E. Eliou, "View from Athens," *NATO's Fifteen Nations* 25, no. 1 (February–March 1980): 44. The plan's preamble stressed that a settlement of Greek-Turkish bilateral issues had to be achieved as a precondition for Greece's reentry into NATO's military structure. The nine articles of the plan altered most of what had been agreed upon between Haig and Davos in mid-1978. While seeking to ensure that Greek territorial interests would not be prejudiced, the Rogers Plan increased the general involvement of NATO units, ostensibly as a neutral element, and widened the range of Turkish activity in the Aegean Sea.

In the parliamentary debates Prime Minister George Rallis, who succeeded Karamanlis after his assumption of the country's presidency in May 1980, and his government ministers stressed that Greece's acquiescence to reentry terms in no way harmed Greek national interests. A "special relationship" provided for fewer Greek forces under direct NATO command in time of peace. Greece would adhere to the long-standing NATO policy of six-mile national airspace for NATO aircraft only. Greek newspapers outlined other general terms of the agreement; for example, the military settlements for Greece's reintegration did not intend to prejudice the outcome of other bilateral disputes between Greece and Turkey. Following the establishment of a new Greek command in Larissa, there would be cooperation with the Turkish command in Izmir for air defense and radar coverage of the Aegean Sea. The two states agreed to begin negotiations to delimit the command and control boundaries. In the meantime, provisional arrangements would remain in force.[24]

The spirit of compromise proved short-lived. Since the acceptance of the Rogers Plan for Greek reintegration into NATO, negligible progress has been made on outstanding bilateral issues. Thus, Greece continues to insist on the prior recognition of the 1974 operations status quo ante before the establishment of Greek-NATO headquarters in Larissa. Turkey has called for the setting up of the Larissa command before the delimitation of operational boundaries in the Aegean. The same discouraging commentary must be made with regard to both the continental shelf controversy and airspace problems. Because crucial claims to sovereignty rights in the Aegean underlie all the divisive issues, both Greece and Turkey avoid even technical agreements that can possibly establish unfortunate precedents for their individual positions. Consequently, Greece has signed the 1982 Convention on the Law of the Sea which reinforces most of its interests in the Aegean dispute, while Turkey has lined up with the much smaller group of nations responding negatively.[25]

NEW DIRECTIONS IN GREEK FOREIGN POLICY

The victory of Papandreou and his Panhellenic Socialist Movement (PASOK) in the 18 October 1981 elections initially did not augur well for the interests of NATO and the United States in Greece. The former Berkeley

---

[24]George Rallis, *Hores efthenis* [Times of responsibility], (Athens: Euroekdotiki, 1983), pp. 95–100; U.S., Congress, Senate, Committee on Foreign Relations, *NATO Today: The Alliance in Evolution*, 97th Cong., 2d sess., 1982, pp. 88–89; C. Z. Sazanides, "New Developments in the Greco-Turkish Dispute over the Aegean Airspace," *Hellenic Review of International Affairs* 2, no. 1 (1981): 120–25; *To Vima*, 21–25 October 1980; Meinardus, "Griechenlands gestörte Verhältnis zur NATO," pp. 109–11.

[25]*UN Chronicle* 19, no. 6 (June 1982): 17. The Turkish representative to the UN Conference noted that some of the convention's provisions could jeopardize the vital legitimate interests of Turkey. *To Vima*, 12 December 1982.

economics professor persistently had harped generally on their negative roles and influence in Greece and specifically on their apparent support of the seven-year military dictatorship and their nonopposition to the Turkish invasion of Cyprus. Stridently nationalistic, Papandreou, as an opposition spokesman, ceaselessly taunted conservative prime ministers Karamanlis and Rallis. His tactics forced them at times to take more rigid positions on foreign policy issues than they might have without such pressures (the Turks also had their own hard-line opposition).

To date, however, Papandreou has not made a sharp shift in Greece's broader relations with NATO and the United States. Indeed, by cutting through the political rhetoric and posturing, one can observe Papandreou extending, albeit with sharper language and different tactics, policies launched by the putatively pro-West and more moderate Karamanlis. Immediately after assuming the prime ministership under troubled circumstances in the summer of 1974, Karamanlis sought to modify Greece's foreign policy path and to shore up the nation's shattered diplomatic status. Part of the objective of developing Greece's freedom to maneuver is mirrored in the decision to withdraw from the military command of NATO. For this bold act Karamanlis could fall back on the example of France.

Subsequent policy showed Karamanlis attempting to steer Greece away from the dominating influence of the United States. His major initiative in this direction is reflected in Greece's intensive campaign to become the tenth member of the European Economic Community (EEC). Karamanlis began this process during an earlier prime ministership when Greece achieved associate status in 1962. This agreement projected Greece's full membership by 1984. Although the EEC had frozen Greece's treaty of association during the seven-year dictatorship, Karamanlis set out to speed up Greece's proposed entry date. Community membership was seen as helping buttress the nation's fledgling democratic institutions and allowing it to downgrade the role of the United States in Greek affairs. Domestic opposition to membership came from the pro-Soviet Communist Party and from PASOK, with Papandreou claiming that Greece would suffer economically and politically. Potential resistance from EEC states for early entry gradually melted away, and formal accession for Greece was completed on 1 January 1981.

While regularly emphasizing that "Greece belongs to the West," Karamanlis embarked on a diplomatic campaign to cultivate relations with regional states. Through these extensive efforts, he sought to take advantage of Greece's geographic position as a crossroads of three continents and as a nation that could provide positive services by a more independent existence. He cogently summed up his intentions by referring to Greece as "the Mediterranean balcony of Europe." Athens thus was trying to reverse a historical pattern by which Greece traditionally suffered for its geopolitical location and in its relations with stronger powers.

Another dimension of Greece's reconstructed foreign policy focused on improving Greece's already favorable relations with the Arab world.

Concurrent with initiatives in the Middle East, Karamanlis also concentrated on bettering Greece's relations with its neighbors to the north and Warsaw Pact states. The spirit of détente in Europe facilitated this process, and the Soviet bloc welcomed these moves. Beyond the benefits of increased trade, improved cordial relations with the Communist states would permit Greece to concentrate on the more immediate threat posed by Turkey. The rewards of this northern rapprochement were reflected in Minister of Defense Evangelos Averoff's remarks to parliament in the fall of 1979 that Greece was in no way threatened by its northern neighbors.

The perception of Turkey as the primary threat to Greek national security influenced Athens's relations with Washington and NATO. In late March 1976 the United States initialed a defense cooperation agreement with Turkey which provided for $1 billion in U.S. military aid for four years as compensation for the use of American military installations on Turkish territory. The Greek government quickly initialed a similar agreement for $700 million over four years in return for the presence of American facilities on Greek soil. Although neither agreement was subsequently implemented, the precedent was established, at Greek insistence, of calculating military aid to the two feuding allies, with the ratio of 7 to 10. Greek arguments have stressed that such a formula will secure a balance of power in the Aegean and discourage Turkish aggression. To ease further Greek apprehensions, Secretary of State Henry Kissinger sent a letter to Greek Foreign Minister Demetrios Bitsios on 10 April 1976. In this communication he stated that Greek-Turkish disputes "must be settled through peaceful procedures and that each side should avoid provocative actions." Moreover, the United States "would actively and unequivocally oppose either side's seeking a military solution and will make a major effort to prevent such a course of action."[26]

Athens subsequently sought to redefine the status of U.S. bases on Greek territory, whose prerogatives, established in 1953, symbolized for many Greeks a colonial presence. Some new guidelines were initialed by the two countries in July 1977, but their implementation became frozen, partly because of Greece's continuing status outside of NATO's military command.[27] Additionally, Athens believed that to sign at this point would encourage Congress to approve an equivalent agreement with Ankara, thereby ending the American arms embargo to Turkey, which finally occurred during the summer of 1978.

[26]The complete text of the letter is in Bitsios, *Pera apo ta synora, 1974–1977*, pp. 253–54. Thanos Veremis, "Greek Security: Issues and Politics," *Adelphi Papers*, no. 179 (London: Institute for Strategic Studies, 1982), pp. 20–21; J. Siotis, "La situation internationale de la Grèce et la demande hellénique d'adhésion aux Communautés," *La Grèce et les Communautés: Problèmes posés par l'adhésion* (Brussels: Editions de l'Université de Bruxelles, 1978), pp. 47–62.

[27]For example, negotiations determined that the four major U.S. bases in Greece—those at Suda Bay and Heraklion in Crete and the Hellenikon Air Base and Nea Makri communication installation in the greater Athens area—would have a Greek commander and require Greek permission for operation during time of war.

Serious negotiations resumed, however, after the United States signed an agreement with Turkey in March 1980. The terms defined the status of American bases on Turkish soil and, of particular importance for Greece, called for significant amounts of military and economic aid to Turkey. It was at this point during the spring that Athens publicly stated that the maintenance of U.S. bases in Greece, which ostensibly served a NATO function, would be tenuous as long as the Greek armed forces remained outside of NATO. Prime Minister Rallis's government rejected American proposals in September 1980, claiming that they reversed many of the terms agreed upon in July 1977. After Greece's reintegration in October 1980, Athens provided counterproposals, essentially reiterating the earlier agreements.

More intensive deliberations began in January 1981. The New Democracy government sought the basic terms of 1977, substantial military aid, support for the development of Greece's nascent arms industry, and American backing for the regional balance of power. Bearing in mind the forthcoming election, the Rallis government naturally desired an agreement that would be favorable for its campaign. American negotiators, probably expecting an electoral reversal for Rallis's party, did not respond with favorable compromises that might provide the sharp focus of volatile criticism before and after the anticipated victory of Papandreou. This round of deliberations came to a formal close with no settlement on 16 June 1981 because of the approaching election.[28]

While never deviating from a general commitment to the West, the rightist governments had succeeded in forging a more flexible foreign policy for Greece in the years following the dictatorship's fall. Criticism of NATO and the United States spanned the broad spectrum of the Greek population from the parliamentary right to the Marxist left, with varying degrees of intensity. During the first days after his October 1981 electoral victory, Papandreou indicated that he would not embark on an adventurous course. That Greece should not be in NATO and that there should be no American bases on Greek territory were presented as more long-range objectives. Greece would not act unilaterally and would avoid a direct challenge to U.S. interests. With his constant focus on the Turkish threat, on the need for a militarily strong and independent Greece, and on past injustices by allies, Papandreou delineated a solidly nationalistic foreign policy that the political right and armed forces would have difficulty criticizing.

Departing from the generally "proper diplomacy" of his predecessors, Papandreou characteristically broadcast shrilly Greece's positions in his attempt to mold a more independent foreign policy within the framework of NATO. Thus, at the December 1981 meeting of NATO ministers, Papandreou prevented the issuance of the customary joint communiqué at the end of the session, largely in response to NATO's decision not to provide the requested

---

[28]Rallis, *Hores efthenis*, pp. 197–98, 203–05, 223–26; *To Vima*, 29 August 1982, and 19 June 1983.

security guarantee for Greece's frontier against potential Turkish aggression. He has regularly stressed that the threat to Greece's security comes not from the Soviet bloc to the north but from Turkey to the east. The PASOK government has recognized as valid the October 1980 Rogers Plan for Greece's reentry into NATO's military structure, but, since delimitation of command boundaries in the Aegean remains an open question, it has to be seen as partially suspended. Greek-Turkish negotiations over outstanding points of difference, hardly productive prior to Papandreou's assumption of power, ceased on Greek initiative in November 1981. The two states agreed to respect a moratorium in July 1982, by which they would abstain from all provocative statements and actions for an indeterminate period. The moratorium proved tenuous, as Athens continued to protest Turkish overflights in the contested six- to ten-mile zone of Greek national airspace. Greek grievances over airspace and operational command problems in the Aegean also have been reflected in Greece's nonparticipation in a number of NATO-sponsored maneuvers in the region. A Cold War atmosphere continues to blanket relations between the two nations. Consequently, their disputes continue to undermine NATO regional security.

Whereas Karamanlis utilized an evenhanded approach to cultivate relations in the Arab world and other regions, Papandreou has outspokenly expressed his support for the Palestine Liberation Organization (PLO) and the Arab cause, liberation movements, and Third World issues. In doing so, he has occasionally blundered by alienating one of the parties in a regional dispute. His granting of diplomatic representation to the PLO and his strong backing for the organization during the Israeli attacks on Beirut have provided few, if any, substantive returns as the Arab states bicker among themselves and Arab oil revenues decline. Indeed, Turkey, with its quiet diplomacy and no history of cordial relations with the Arab world, has augmented its stature among the Arabs in recent years and benefited economically.

Papandreou also has extended more boldly Greece's contacts with Soviet bloc states. A number of his government's policy statements have run counter to the general positions of fellow NATO-EEC states. Athens thus opposed NATO sanctions on Poland after the establishment of the military government there and muted criticism of the Soviet presence in Afghanistan. Moreover, a number of Papandreou's statements have supported Soviet policy positions. At the conclusion of Soviet Prime Minister Nikolay A. Tikhonov's visit to Athens in February 1983, which resulted in a ten-year economic accord, a joint communiqué emphasized that nuclear-free zones were an important part of the disarmament process. Papandreou has publicly recorded his opposition to the deployment of *Pershing II* and cruise missiles and expressed the resolve of his government to set up a nuclear-free zone in the Balkans.[29]

[29]*New York Times*, 25 February 1983; *Christian Science Monitor*, 31 March 1983; *International Herald Tribune*, 29 July 1983.

Despite these obstreperous forays away from established NATO positions, Papandreou has not broken Greece's long-standing links with the West. Rather than take Greece out of the EEC, as he had threatened, he has instead called for a rewriting of the terms of participation in order to bolster Greece's highly vulnerable economy. Indicative of an attitude to work within the system, Greece assumed the rotating six-month presidency of the EEC on 1 July 1983, with requisite propriety and considerable internal publicity.

Perhaps a more significant sign of a basically unchanged orientation is to be found in the initialing of the Defense and Cooperation Agreement between Greece and the United States on 15 July 1983. The difficult negotiations for the revision of the 1953 U.S. bases accord had begun in October 1982 and were accompanied by public signs of uncertainy on both sides. Ultimately, the negotiating objectives of the PASOK government adhered generally to those pursued by its conservative predecessors, except for the attention placed on a timetable for the closing of the bases. Papandreou has emphasized that the dismantling process of the bases will begin after five years, starting on 31 December 1988. The American interpretation of the vague wording focuses on the opportunity to extend the longevity of the installations after that date. Greece has received a generally worded American commitment to maintain the balance of military power between Greece and Turkey in the Aegean. It is projected that Greece will be the recipient of $500 million in military aid during 1984, a sum faithful to the 7-to-10 Greek-Turkish aid ratio asked for by the earlier Karamanlis and Rallis ministries. The Reagan administration will not commit itself, however, to this restrictive formula as a permanent guideline. Greece also will have the right to react to the use of these bases against third countries, particularly those which are considered friendly to Athens. Additionally, the irksome extraterritoriality privileges of American personnel, which prevented those individuals committing crimes and misdemeanors from being tried in Greek courts, will be modified on Greece's behalf. The anticipated purchase of large amounts of sophisticated weaponry and aircraft from the United States will continue a military dependency on Western sources.[30]

Papandreou's government has yielded to certain harsh realities. A falling-out with the United States and, consequently, with NATO could only improve Turkey's diplomatic and military muscle in the Aegean and Cyprus and accentuate Greek weaknesses. The United States has consented to maintain a military balance between the two feuding allies, thereby providing the support that Greece could not expect to find from other sources. Papandreou has cogently declared that Greece's relations with the United States and NATO pass through Ankara.

Interestingly enough, the reports of TASS and Radio Moscow indicated approval of the Greek position and accomplishments on the base agreement,

[30] *International Herald Tribune*, 16–17 July 1983; *Kathimerini* (Athens), 16 July 1983; *To Vima*, 17, 24 July 1983.

citing that Athens's firm stand forced the United States to make certain concessions and that the American military in Greece would be severely limited in its activities. Speculation on Soviet motives center around Moscow's apparent desire that the balance of power in the area of the Mediterranean not be markedly disturbed. Moreover, a Papandreou-led Greece within NATO can promote policies favorable to Warsaw Pact interests, as has recently been the case.[31]

Papandreou, in turn, earlier acknowledged forces beyond Greece's ability to control by stating, for example, during his BBC interview of 26 October 1981: "We're not prepared to accept that Potsdam and Yalta must still hold and determine the life of Europeans. We don't believe in the two armed camps, Warsaw Pact and NATO."[32] The political decisions of 1944–45 do still seem to hold, however. Indeed, Greece since its independence has had to deal with and been forced to consider as "friends" those powers that are dominant in the Mediterranean. Although there is a significant Soviet presence in the Mediterranean, its northern rim is dominated by NATO states and its waters are patrolled by large units of NATO nations.

The at times contradictory perceptions of the Papandreou government are evidence of its attempts to walk the diplomatic tightrope toward the goal of what it refers to as a "national, independent, and multidimensional foreign policy." For that matter, the sum total of Papandreou's initiatives within NATO appears to presage a role for Greece similar to that of Romania in the Warsaw Pact. The reasons and circumstances for such roles are obviously different, but, just as there are realistic limits within which Romania has maneuvered independently of Soviet policy since the 1960s, there also seem to be restrictive forces on the PASOK government which keep it from drifting into more adventurous diplomacy. In the meantime, Papandreou's Socialist regime will no doubt continue to create uncomfortable moments for Greece's alliance partners.

---

[31]*International Herald Tribune*, 29 July 1983.

[32]*Multidimensional Greek Foreign Policy: Interviews of Prime Minister Andreas G. Papandreou over ABC and BBC 1* (Athens: General Secretariat for Press and Information, 1982), p. 18.

Turkey and NATO

BRUCE R. KUNIHOLM

Thirty years after Turkey joined the North Atlantic Treaty Organization many of the geopolitical realities that provided the context for its accession still hold. Shifts in the international balance of power and crises of confidence and trust have created strains in the alliance, but the bonds between Turkey and its allies are nonetheless secure. New developments, meanwhile, have given Turkey added importance in the geopolitical picture. Strategic parity between East and West and a changed strategic environment in the Near East together make Soviet adventurism somewhat more threatening than it was earlier; greatly increased Soviet naval and air power makes the Soviet Union a much more important player in the eastern Mediterranean; and the increased importance and vulnerability of Persian Gulf oil, which in 1952 provided approximately one-tenth and which now provides over one-half of Europe's oil, have made the Persian Gulf region vital to the security of NATO. As a result, while Turkey's part in the defense of Europe—and Turkey in Asia—continues to be the focal point of its security concerns, its potential role in the security of the Middle East and Persian Gulf has become increasingly significant. The range of geopolitical and other factors that have contributed to these developments will be explored here, and their implications examined for the future of Turkey's relationship with NATO.

THE EARLY YEARS

When NATO deliberated over Turkey's accession to its ranks in 1950–51, there were differences between those American officials who gave priority to the balance of power in Western Europe and those who argued that the balance of power in the Near East was almost, if not equally, important. But with the advent of the Korean War and implementation in the U.S. government of a policy that seemed to show that distinctions between peripheral and vital interests were unnecessary, categorical geographical conceptions about Turkey were altered. Turkey's unique strategic position, enhanced by the changed face of postwar geopolitical realities, gave it an important role to play in the defense of both regions.[1]

[1]For a detailed examination and documentation of the argument put forward in the first part of this essay, see Bruce Kuniholm, "The Near East Connection: Greece and Turkey in the Reconstruction and Security of Europe" (Paper delivered at the symposium "U.S.-European Relations, 1945–1952," Institut für Geschichte, University of Salzburg, Austria, 17 April 1983).

A prerequisite for Turkey's accession to NATO, however, was a sense of reciprocity; only a conviction that there was a mutuality of benefits could make credible, and therefore possible, a mutuality of obligations. By late 1951 this requirement had been met. Support for the balance of power in the Near East interlocked with support for the balance of power in Europe. Geographic distinctions between the two regions were dismissed as artificial, and the two came to be regarded as interconnected, with Turkey serving as the linchpin.

The rationale behind NATO's decision was interesting for what it revealed about American perceptions of Turkey and how they had changed since Soviet activities in the Near East in 1946 had forced the United States to focus on Turkey. In what President Harry S. Truman regarded as his most important decision since Hiroshima, he decided in August 1946 that the United States would resist with all means at its disposal, including the force of American arms, any Soviet aggression against Turkey. This decision laid the groundwork for the Truman Doctrine in 1947.

In 1950–51 Turkey was still seen as the strongest anti-Communist country on the periphery of the USSR. The only nation in the eastern Mediterranean capable of substantial resistance to the Soviets, it constituted a deterrent to aggression and provided something of a protective screen for the region. Loss of Turkey to Soviet influence would give the USSR a valuable strategic position in the region. If the Soviets attacked Iran, and Turkey were only neutral, the Soviet right flank would be protected. Oil interests in the Gulf, meanwhile, would be vulnerable, and European economic viability would be threatened. None of this was new.

What was new in the early 1950s was the conviction of a mutuality of benefits; Turkey and Greece, on the one hand, and Europe and the United States, on the other, all stood to benefit. Turkey had been linked by a treaty of mutual assistance with Britain and France since 19 October 1939. The treaty had last been reaffirmed in 1949, although no one was sure how much emphasis to place on it. Since the United States believed that the Soviets would attack Turkey only if they were prepared to face the prospect of global war, and since the United States had profound political and strategic interests in Turkey, as well as a heavy moral commitment to it, many officials believed that the United States would assist the Turks whether or not it had a reciprocal treaty arrangement with them. In effect, the United States already was committed to the security of Turkey. For the United States to obtain Turkey's full cooperation in order to maximize its contribution in international security issues, or to assure its cobelligerency in the event of an attack on Europe, a U.S. security commitment was required. Such a commitment was also necessary to secure access to its valuable bases and to close the Straits at the first sign of hostilities.

Without a formal security commitment to Turkey, the United States was concerned that the government would drift toward neutrality, as it had in World War II, and as Iran appeared to be drifting at the time under Premier

Mossadegh. If this happened, officials reasoned, the United States would lose the assistance of a powerful ally. Greece and Iran, if they were to be protected, required mutual defense arrangements with Turkey. This was only possible under a broader security arrangement, of which the United States was a part.

As members of NATO, Turkey and Greece would be important to SACEUR, both as a deterrent to a Soviet attack and as a threat to the Soviet Union's southern flank. If these regions' military potential were integrated in a security framework, the USSR would have to commit significant forces to protect itself in the southwest and to safeguard its vital oil fields around Baku. A security commitment to Turkey, therefore, would constitute a far more effective deterrent than previous arrangements against a Soviet attack, not only along the Middle East's entire northern tier, which provided a buffer for European and American oil interests in the Persian Gulf, but also in Europe.

The Turks believed that they badly needed this commitment. Experience had taught them the value of a credible deterrent. They had, moreover, something to contribute. As President Celâl Bayar told Assistant Secretary of State George McGhee, Turkey "wants to give a guarantee, and it would like to receive a guarantee."[2] Stalin, Ankara knew, harbored little love for the Turks. That there was substance to Turkish concerns is evident both in Stalin's attempts to acquire portions of eastern Turkey and the Straits in 1945–46 and in his successors' mea culpa less than three months after his demise. In a 30 May 1953 note the Soviets informed the Turks that the governments of Armenia and Georgia had renounced their territorial claims against Turkey. They also stated that, after reconsidering the question of the Straits, they believed Soviet security could be assured by conditions acceptable to Turkey, an unusual public retraction and tacit admission of Stalin's irresponsible international ambitions.[3] By then, however, Turkey and Greece were members of NATO, and Soviet attempts to alter Turkey's alliance relationship fell on deaf ears.

THE MIDDLE YEARS

The golden era of Turkey's association with NATO in the 1950s gave way in the 1960s and 1970s to a more complicated relationship. Strains between the NATO allies were precipitated by gradual shifts in the international balance of power, represented by the Cuban missile crisis of October

[2]U.S., Department of State, *Papers Relating to the Foreign Relations of the United States, 1951* 3 (Washington, DC: Government Printing Office, 1981): 470.

[3]Ferenc Váli, *Bridge Across the Bosporus: The Foreign Policy of Turkey* (Baltimore: Johns Hopkins, 1971), pp. 174–75. For Nikita Khrushchev's recollections of Stalin's conduct, see Bruce Kuniholm, *The Origins of the Cold War in the Near East: Great Power Conflict and Diplomacy in Iran, Turkey, and Greece* (Princeton: Princeton University Press, 1980), pp. 358–59.

1962, and by crises of confidence and trust, represented by differences between the United States and Turkey over Cyprus.

The Cuban missile crisis and the Kennedy administration's abandonment of a strategy of massive retaliation for one of flexible response together constituted a watershed in the Cold War.[4] In Turkey, however, these developments were contributed to a concern that, in the event of Soviet aggression, Turkish territory might be traded for time, and that NATO would favor its central front over its flanks, defending the former at its perimeter and the latter only in depth or not at all.[5] The announcement in January 1963, only three months after the missile crisis, that *Jupiter* missiles would be removed from Turkey (they were out by mid-April) had a significant effect on the Turks.[6] Along with subsequent revelations that a bargain had been struck, the Cuban missile crisis emphasized Turkey's vulnerability to unilateral decisions by the United States.[7]

Turkey's vulnerability was further emphasized during the Cyprus crisis of 1963–64, when President Lyndon B. Johnson warned Prime Minister Ismet Inönü that Turkey should not use U.S. supplied military equipment to invade Cyprus. In a letter of 5 June 1964 the president informed Inönü that, if the Turks took a step that resulted in Soviet intervention, Turkey's NATO allies might have to reconsider their obligation to protect Turkey against a Soviet attack.[8] By calling into question U.S. obligations under NATO, the Johnson letter raised serious doubts among the Turks about the reliability of the United States as an ally. These doubts, in turn, produced a more realistic assessment of Turkey's problems and a more independent, multifaceted conception of Turkey's options than that which had been possible in the early Cold War years. Such a conception was encouraged by Soviet President Nikolai Podgorny, whose visit to Turkey in January 1965 played on an earlier theme

---

[4]Alvin Rubinstein, *Soviet Policy Toward Turkey, Iran, and Afghanistan: The Dynamics of Influence* (New York: Praeger, 1982), pp. 15–19.

[5]George Harris, *The Troubled Alliance: Turkish-American Problems in Historical Perspective, 1945–1971* (Washington, DC: American Enterprise Institute and the Hoover Institution on War, Revolution and Peace, 1971), pp. 93–95; Váli, *Bridge Across the Bosporus,* pp. 120–21; Ihsan Gürkan, "NATO, Turkey, and the Southern Flank: A Mideastern Perspective," Agenda Paper no. 11 (New York: National Strategy Information Center, 1980), pp. 52–53. For more recent expression of these concerns stemming from the allocation of AWACS within NATO, see Gürkan, "NATO, Turkey, and the Southern Flank," pp. 36–37.

[6]Donald Hafner, "Bureaucratic Politics and 'Those Frigging Missiles': JFK, Cuba, and US Missiles in Turkey," *Orbis* 21, no. 2 (Summer 1977): 319; Váli, *Bridge Across the Bosporus*, p. 132; Rubinstein, *Soviet Policy*, p. 19; George Harris, "The View from Ankara," *The Wilson Quarterly* 6, no. 5 (Special Issue 1982): 132.

[7]Duygu Bazoğlu Sezer, "Turkey's Security Policies," *Adelphi Papers*, no. 164 (Spring 1981): 26–27; Váli, *Bridge Across the Bosporus*, p. 132; Robert Kennedy, *Thirteen Days: A Memoir of the Cuban Missile Crisis* (New York: W. W. Norton, 1969), pp. 108–09.

[8]The Johnson letter is reprinted along with Inönü's reply in *Middle East Journal* 35 (Summer 1966): 386–93.

sounded by Stalin's successors. This time, Soviet overtures found a far more receptive audience.[9]

On 5 January 1965, Podgorny told the Turkish Grand National Assembly that for some time after World War II a shadow had been cast over Turco-Soviet relations and acknowledged that inappropriate and incorrect statements made in the Soviet Union had played a negative part. The Soviet Union, he noted, declared openly that those statements were not correct; such events should be a part of the past. Podgorny's comments came during the first visit to Turkey by a Soviet parliamentary delegation in more than twenty-five years. They were seen in Turkey as an indirect admission of Soviet responsibility for strains in relations with Turkey in the late 1940s and in the 1950s, and they paved the way for better relations between the two countries.[10]

Improvement in Turco-Soviet relations was marked by visits, principled agreements, and economic assistance and complemented a diminishing level of grant assistance from a United States increasingly mired in the war in Vietnam. Revenues also were supplemented by remittances from Turkish workers in Germany. Freedom from the ideological straitjacket that had characterized early postwar policies resulted in a somewhat more independent international posture; for example, the decision in 1965 not to participate in the Multilateral Nuclear Force.[11]

While the Turks continued to distrust the Soviets, their policies clearly were more flexible and less structured by the earlier assumptions of a bipolar world. The new balance that was evident in Turkish policy could be seen by comparison with past events. During the civil war in Lebanon in 1958, for example, the United States had used Turkish bases to support its intervention. Following events of the early 1960s, the Turks were more guarded about the use of their facilities for non-NATO contingencies. During the 1967 Arab-Israeli War, the United States was allowed to use communication stations in Turkey but was not given permission to use Turkish bases for refueling or supply activities. In the course of the 1973 Arab-Israeli War, the United States was prevented from using Turkish bases for direct combat or logistical support, although it was allowed to use communication stations in Turkey during the resupply effort. The United States also could utilize Turkish bases for the

---

[9]Francis P. Butler, "Reassuring Turkey: A Faithful Ally Disillusioned and in Trouble," *Parameters* 10, no. 1 (March 1980): 24–32; Rubinstein, *Soviet Policy*, p. 20; Harris, *Troubled Alliance*, pp. 105–28; Harris, "View from Ankara," p. 135; Váli, *Bridge Across the Bosporus*, pp. 128–34, 176–81; Sezer, "Turkey's Security Policies," pp. 23–24, 30; Andrew Mango, "Turkey: A Delicately Poised Ally," *The Washington Papers* 3, no. 28 (Beverly Hills, CA: Sage Publications, 1975): 34.

[10]See also the statement by Podgorny at Izmir, 11 January 1965. Foreign Broadcast Information Service, *Daily Report*, no. 3, (6 January 1965): M1–3; no. 4 (7 January 1965): M1–5; no. 6 (11 January 1965): M1–4.

[11]Sezer, "Turkey's Security Policies," pp. 23–24, 32; Harris, "View from Ankara," pp. 133–34.

evacuation of American citizens during the civil war in Jordan in 1970 and the Iranian Revolution in 1979.[12]

THE RECENT PAST: TURKEY IN CRISIS

The Cyprus crisis in 1974 again tested mutual obligations and responsibilities. The purpose of Turkish military intervention in July, justified by Turkey's obligation under the 1960 Treaty of Guarantee, was to protect the Turkish minority from the "Hellenic Republic of Cyprus" which was led by an international terrorist, installed by a coup, backed by the military dictatorship in Athens, and bent on union with Greece. A second Turkish action in August—to consolidate vulnerable positions according to the Turks and to expand their base according to the Greeks—precipitated a U.S. congressional embargo on transfers of military equipment to Turkey, effective 5 February 1975, and resulted in a subsequent decision in Ankara to suspend U.S. operations at military installations in Turkey, as of 26 July 1975. These developments made explicit what had been implicit until then: access to facilities was directly related to decisions on military assistance. Blunt assessments of the problem called into question fundamental assumptions on both sides and forced discussion over the merits of continuing the special relationship.[13]

Relations between Ankara and Moscow, meanwhile, continued to improve. By 1978 the Soviet Union was aiding forty-four different development projects in Turkey, and by the end of the decade Turkey received more Soviet economic assistance than any country in the Third World. If Prime Minister Bülent Ecevit could declare in May 1978 that Turkey felt "no threat" from the Soviet Union, that statement, however required qualification, particularly after the Soviet invasion of Afghanistan, which worried the Turks far more than had the fall of the shah.[14]

---

[12]U.S., Congress, House Committee on Foreign Affairs, *Turkey's Problems and Prospects: Implications for U.S. Interests*, 96th Cong., 2d sess., 1980, pp. 11, 48; Rubinstein, *Soviet Policy*, p. 38.

[13]U.S., Congress, House, Foreign Affairs Committee, *Congressional-Executive Relations and the Turkish Arms Embargo*, Congress and Foreign Policy Series, no. 3, 97th Cong., 1st sess., June 1981, p. 27. (This study is especially useful in illuminating the role played by Watergate and the Vietnam War.) Harris, *The Troubled Alliance*, p. 208; Rubinstein, *Soviet Policy*, p. 46; Gürkan, "NATO, Turkey, and the Southern Flank," p. 38. As one analyst has pointed out, the Turks saw as hypocritical the U.S. focus on the origins of the arms used by Turkey rather than on the rationale for their use. Butler, "Reassuring Turkey," p. 28. Selective application of U.S. arms transfers law (the U.S. Foreign Assistance Act and the Foreign Military Sales Act) against Turkish use of American arms in Cyprus is highlighted by congressional failure to apply those laws with equal force to Israeli use of American arms in Lebanon. The embargo was lifted on 26 September 1978; as a consequence, Turkey allowed the United States to resume operations at military facilities on 9 October.

[14]George E. Gruen, "Ambivalence in the Alliance: U.S. Interests in the Middle East and the Evolution of Turkish Foreign Policy," *Orbis* 24, no. 2 (Summer 1980): 371. The Soviet Union was also the most important source of assistance for Turkey's public sector projects.

Additional cause for concern grew out of political and economic problems manifested by systemic complications and endemic civil violence. By September 1980 an average of over twenty politically motivated murders were being committed in Turkey every day, and more than thirteen thousand people had been killed or wounded over the previous two years (numbers differ depending on the definition of motives involved). Two-fifths of Turkey's work force were unemployed, and the rate of inflation had reached 128 percent.[15]

Political Factors

Explanations for these developments vary. Some analysts have focused on the constitution of 1961, drawn up by Turkey's leading intellectuals to correct the excesses of Prime Minister Adnan Menderes during the 1950s and to address in particular his regime's failure to promote social justice. In seeking to prevent a repetition of the previous regime's abuses, however, the intellectuals created a system of proportional representation that encouraged political fragmentation and inadvertently impeded the legitimate exercise of power by subsequent regimes. By the 1970s, in spite of threats by the military to intervene if reforms did not address rising domestic instability and strife, neither the Justice Party nor the Republican People's Party, which together were supported by 80 percent of the electorate, was able to rule with a majority. The parliamentary system could not—or would not—produce either the stable majorities or compromises necessary for effective political action. As a result, Turkey was governed by twelve minority or coalition governments. In this context, small factions were able to exploit religious or ethnic differences within Turkey, thus exercising influence out of proportion to their numbers. Politics by coalition gradually degenerated into what one analyst has characterized as politics by deadlock.[16]

Whether the blame for Turkey's difficulties should be placed more on the constitution than on the politicians who operated under its law is a moot point. Some Turkish professors and journalists stand by the 1961 constitution and find fault with Turkey's previous political leaders. According to one

Rubinstein, *Soviet Policy*, pp. 27, 40. See also Sezer, "Turkey's Security Policies," pp. 27–29, 33, 40–43; *New York Times*, 30 May 1978; and Bülent Ecevit's address to the International Institute of Strategic Studies, 15 May 1978, in *Survival* 20, no. 5 (1978): 203–08.

[15]Nicholas S. Ludington, and James W. Spain, "Dateline Turkey: The Case for Patience," *Foreign Policy*, no. 50 (Spring 1983): 154–56; *Turkey, Greece, and NATO: The Strained Alliance* (Staff report to Committee on Foreign Relations [Washington, DC: Government Printing Office, 1980], p. 22).

[16]George Harris, "The Left in Turkey," *Problems of Communism* 29, no. 4 (July-August 1980): 26–28; Paul Henze, "On the Rebound," *The Wilson Quarterly* 6, no. 5 (Special Issue 1982): 120; U.S., Congress, Senate, Subcommittee on Security and Terrorism of the Committee on the Judiciary, *Hearing on Terrorism: The Turkish Experience*, 97th Cong., 1st sess., 25 June 1981, p. 18; *Turkey, Greece, and NATO*, pp. 22–24; Maxwell O. Johnson, "The Role of the Military in Turkish Politics," *Air University Review* 33 (January-February 1982): 57; Sezer, "Turkey's Security Policies," pp. 8, 20; *Washington Post*, 20 November 1982.

observer of its political scene, Turkey's main political parties were networks of interests concerned with patronage; they were led by individuals who gave immediate satisfaction to material wants, thereby sacrificing short-term advantage for long-term improvement. As a result, they neither implemented needed structural reforms nor created the indigenous institutions required to regulate political freedom. Because of their personal failings and lack of prudence, they magnified systemic problems.[17] In conjunction with terrorism, which began to ravage Turkey in the late 1970s, these systemic problems placed the country on the verge of a civil war.

Economic Explanations

If some analysts of the crisis in Turkey underscore the role of political variables, others point to economic issues to explain Ankara's difficulties, particularly the lack of developed indigenous energy resources and the costs of energy imports. Turkey was particularly vulnerable to the enormous increases in oil prices in 1974 and again in 1978–79. Its oil bill, which was only $124 million in 1972, rose to $1.2 billion in 1977 and to $3.86 billion in 1980. These staggering costs and a recession in Europe, which in the mid-1970s resulted in the reduction of Turkey's important annual income from remittances, together were central to Turkey's difficulties. When economic problems were compounded by bad management, inefficient government-run institutions, and a lack of firm political leadership, they explain why Turkey's external debt rose from $2.8 billion in 1973 to $17 billion by 1979. The oil bill, which in 1980 was approximately equal to Turkey's $3.7 billion balance of payments deficit, ate up foreign exchange earnings. After 1976 increasing deficits, financed initially by a draw down in reserves, were financed almost entirely by building up the external debt.[18]

One analyst, who acknowledges the difficulties associated with the cost of imported oil, has looked hard at Turkey's internal economic problems and regards the difficulties examined here as stemming neither from a political crisis caused by an unworkable political system, nor from an economic crisis created by external factors, which are, nonetheless, important. Rather, he sees them as "the culmination of a fundamentally flawed development strategy."[19] Mehmet argues that the Kemalist experiment in étatism (state

[17]Andrew Mango, "Understanding Turkey," *Middle Eastern Studies* 18, no. 2 (April 1982): 201–02.

[18]Organization for Economic Cooperation and Development surveys, *Turkey* (Paris: OECD, 1982), pp. 7, 43; Sezer, "Turkey's Security Policies," p. 4; *Perspectives on NATO's Southern Flank* (Report to Committee on Foreign Relations [Washington, DC: Government Printing Office, 1980], pp. 8–9); *Wall Street Journal,* 17 September 1982.

[19]Özay Mehmet, "Turkey in Crisis: Some Considerations in the Kemalist Development Strategy," *International Journal of Middle East Studies* 15 (1983): 47–66. Mehmet argues that there may be good historical reasons for supporting étatism and the objective of self-reliance when the two are pursued separately, but that the two together reinforce the growth of a non-competitive, closed economy. He acknowledges that unregulated market capitalism may generate

partnership, ownership, and planning of the economy) has become increasingly doctrinaire. Intended at first to initiate and develop projects in fields in which private capital was inadequate, étatism in fact has promoted monopoly, discouraged market competition, and in effect brought about a noncompetitive closed economy dominated by badly managed, inefficient state economic enterprises that have been losing approximately $2 billion per year. As a result, Turkey was unable to respond creatively to the economic crisis that it faced as the decade drew to a close.

The government of Suleyman Demirel, sensitive to such arguments and increasingly responsive to recommendations made by the OECD, the IMF, and the IBRD, began to address some of Turkey's economic problems by introducing on 24 January 1980 the first of a series of broad-based economic stabilization measures that were continued after the generals took over. Placing great reliance on market forces, the Turkish government eliminated subsidies to inefficient public sector enterprises, curtailed imports, increased exports by devaluing the lira, cut oil consumption, introduced a tight monetary policy to limit inflation, and removed barriers to foreign investment.[20]

In time, these economic measures would turn around the economy of Turkey. The rate of oil consumption was reduced (oil imports, which cost $3.862 billion in 1980, peaked at $3.878 billion in 1981 and were reduced to $3.64 billion in 1982). Exports were increased from $2.9 billion in 1980 to $4.7 billion in 1981, when processed and manufactured products ($2.29 billion) exceeded agricultural products ($2.22 billion) for the first time, and to $5.75 billion in 1982. Inflation, at 94% in 1980, was reduced markedly to 38% in 1981, to below 30% in 1982, and to 25% in the first quarter of 1983. At the time these reforms were instituted, however, social unrest was getting out of control. The dismal economic and political situation created an environment conducive to extremism and terror, and terrorism in turn served to undermine the government's initial attempts at economic reform.[21]

The "September 12 Action" and Its Aftermath

In September 1980, as the economic and political situation grew worse and the death rate by terror reached twenty-eight per day, General Kenan Evren, former commanding officer of the Turkish brigade that fought valiantly in Korea and chief of staff of the armed forces, announced that the military leadership, acting in its traditional capacity as guardian of the state, was

---

large inequalities but asserts that Turkey's previous economy is no guarantee against such a result. Faced with a choice, the Turks would choose a more open, competitive, and decentralized economy.

[20]*Christian Science Monitor*, 23 November 1982; Harris, "View from Ankara," p. 134; *Perspectives on NATO's Southern Flank*, p. 9; *New York Times*, 21 November 1982.

[21]*Turkey*, p. 7; OECD survey, *Turkey* (Paris: OECD 1983), pp. 26, 28, 29, 41; *Terrorism: The Turkish Experience*, pp. 10, 24.

taking over control of the government. The sense of relief with which Turks greeted the restoration of order seemed virtually unanimous.[22]

The generals, meanwhile, pursued a series of bold and increasingly controversial initiatives. Following the military takeover and the restoration of order in Turkey, Evren struck another blow at abuses which he believed to be at the source of Turkey's predicament. A new constitution, drafted to replace the "liberal" constitution of 1961, was submitted on 7 November 1982 to a national referendum. While more controversial than the 12 September action, and in spite of criticism that eligible voters, if they failed to register, would be subject to fines or jail; that voting was mandatory; that Evren ran unopposed; and that there were some instances of intimidation at the polls, most Turks from all walks of life agreed that the results were what the country wanted. The statistics themselves are overwhelming: of 20,690,914 registered voters, 18,885,488 (91.27%) cast 18,841,990 valid votes at 82,913 polling stations. Of the valid votes, 17,215,559 (91.37%) approved of the new constitution and 1,626,431 (8.63%) did not.[23]

In supporting the referendum, the Turks granted Evren broad executive powers, restricted the powers of the legislature and the judiciary, accepted limitations on human rights and liberties, and in effect said "no" to terrorism and "yes" to authority. By the spring of 1983 the generals had struck a third and even more controversial blow against key institutions, purging the universities, tightening their grip on the press, and cracking down on labor unions.[24] While the government's actions clearly angered and concerned Turkey's professional elite and particularly its journalists, they did not elicit the same reaction from the lower and middle classes, many of whom continued to view public safety as a prerequisite to orderly government and the exercise of individual liberties.

---

[22]Michael Turner, "Politics and the Future of Turkish Democracy," *Turkish Studies Association Bulletin* 5, no. 2 (September 1981): 1–4; Henze, "On the Rebound," p. 124; *International Herald Tribune*, 27 April 1983; Chris Kutschera, "The Price of Security," *The Middle East* (September 1981): 32–33. For the traditional historical role, see Johnson, "Role of the Military in Turkish Politics," pp. 49–63. For discussion of the causes of terrorism, see Harris, "The Left in Turkey," *Terrorism: The Turkish Experience;* and Paul Henze, *Goal: Destabilization, Soviet Agitational Propaganda, Instability and Terrorism in NATO South* (Marina del Rey, CA: European American Institute for Security Research, 1981), pp. 15–47. It is worth noting that within ten months of the military takeover the government confiscated 473,553 firearms and that an additional 160,170 were delivered voluntarily, for a total of 633,723. Many of these appear to have been counterfeit; for example, manufactured under a false license in Bulgaria or for a Bulgarian state agency in Argentina. Turkish authorities estimate their cost at $250 million and that of financing terrorist activity from 1977 to 1980 at three times the cost of the weapons—another $750 million.

[23]*The Constitution of the Republic of Turkey* (Ankara: Directorate General of Press and Information, 1982); *Washington Post*, 6, 8 November 1982;*Christian Science Monitor*, 21 July and 9 November 1982; *New York Times*, 9 November and 12 November 1982; *International Herald Tribune*, 27 April 1983.

[24]*New York Times*, 16 March 1981, 17 January, 17 March, 5 April, 3, 24 December 1982, and 6 January 1983; *Washington Post*, 25 April 1983.

As the political environment stabilized under the rule of the generals, the overall economic situation improved, benefiting not only from political stability but also from a continuation of the economic reforms that had been initiated before the takeover. The arrest and imprisonment of thousands of suspected terrorists, however, accompanied as they were by violations of human rights, resulted in widespread international criticism of Turkish policies, criticism which, under the circumstances, many Turks considered misplaced. There was also resentment of calls for political and economic pressure on the government to be more responsive to democratic principles.[25]

Part of the Turkish reaction undoubtedly related to a sense that the international community was neither fully aware of Turkey's recent difficulties nor appreciative of the role traditionally played by the military in Turkish politics. Part of the reaction also may have been conditioned by a general sensitivity to criticism directed at Turkey by ill-informed observers. Whatever the psychological factors involved, one of the reactions of the Evren government to criticism by the West has been to look increasingly to its relationships in the Middle East. In May 1983, for example, General Evren was critical of what he referred to as "Byzantine intrigues" to expel Turkey from the Council of Europe, to restrict Turkish imports, and to criticize its human rights record. Turkey could do without the West, he said, if the West continued to make things difficult.[26] Where could it turn? One direction was indicated a year earlier when Evren declared that Turkey was a European country and, at the same time, a Middle Eastern country.[27] If the Turks' earlier desires to reject their Middle Eastern identity and be accepted as a member of the European Community were now being qualified, economic developments, far

[25]*New York Times*, 16 January, 15 April, 24 May 1981, 14 January, 10 February 1982, and especially 2, 10 July 1982. For Turkish resentment, see, for example, *New York Times*, 5 May 1981, and 17 February 1982. For a list of some of the journalists, writers, actors, and academics who were reported detained and sentenced by the martial law authorities, see Tosum Ersan, "Turkey's Battered Democracy," *Index on Censorship* 1/82, vol. 11, no. 1 (February 1982): 11ff. In April 1983 published official statistics indicated that over the last few years 29,940 persons have been sentenced to prison terms of five years to life, and that 23 of 114 sentenced to death had been executed. Government figures also showed that in the two years since the coup 605 complaints of torture have been investigated, with 177 dismissed as groundless, 368 still under investigation, 47 still in the courts, and 13 resolved. Fifteen deaths by torture have been acknowledged by the Evren government, which has convicted 12 persons of torture, the first administration to do so since the establishment of the Turkish Republic. In June 1983 the generals approved the creation of civilian courts, staffed by civilian judges and prosecutors, to relieve the military tribunals of the more than 61,000 cases involving terrorists which had swamped the courts since the military crackdown on terrorism (close to 30,000 are awaiting trial). *Washington Post*, 6 November 1982, and 25 April 1983; *New York Times*, 17 March and 15 May 1982; *Christian Science Monitor*, 5 April 1983; *International Herald Tribune*, 27 April and 18–19 June 1983; Ludington and Spain, "Dateline Turkey," p. 165.

[26]*International Herald Tribune*, 16 May 1983. For earlier indications of Turkey's desires to develop a "multidimensional, balanced foreign policy," see *New York Times*, 28 November 1982; and *Christian Science Monitor*, 22 December 1982.

[27]Henze, "On the Rebound," p. 125.

more than differences within the alliance and Turkish sensitivity to criticism, suggest why.

From 1980 to 1982, while the percentage of Turkey's exports to the European Economic Community dropped from 42.7% to 30.5% of total exports, the percentage that went to the Middle East and North Africa more than doubled, from 22.3% to 45.0% of the total.[28] In 1982 alone Turkey's construction contracts in the Middle East and North Africa amounted to over $15 billion, including about $8 billion with Libya, $5.1 billion with Saudi Arabia, and $1.15 billion with Iraq. By the end of 1982 there were approximately 250,000 Turkish workers in the region, primarily in Libya (100,000) and Saudi Arabia (54,000).[29]

Turkey's relations with Iran are increasingly important. In early 1982 Iran and Turkey signed a major barter arrangement (agricultural products and manufactured goods in exchange for crude oil). By the end of 1982 Iran was Turkey's major export market. In April 1983 the two countries signed a protocol that called for an increase in volume of trade from close to $1.6 billion in 1982 to $2.5 billion by the end of 1983.[30]

Turkey's relations with Iraq are also significant, particularly for Iraq. The 1,000-kilometer pipeline through Turkey, in operation since 1977, is Iraq's only means of exporting oil while both the Shatt al-Arab waterway and the pipeline through Syria are closed. The pipeline's 650,000-barrel capacity, increased 10 percent by the use of chemicals, is in the process of being enlarged to 1 million barrels per day. One analyst has estimated that in 1982 approximately 6 million tons of goods crossed over from Turkey into Iraq (3 million tons) and Iran (another 3 million tons). That is only 2 million tons less than all the lend-lease goods that went to the USSR through the Persian corridor between 1941 and 1945. The volume of Turkey's trade with Iraq, meanwhile, is second only to that with Germany.[31]

Turks, who pride themselves on their country's European orientation and seek to foster in Turkey some of Europe's social democratic ideals, occasionally have expressed concern over their country's increasingly close association with the Middle East and the greater visibility of Islam in Turkey. They worry about the implications of rural—that is, religious—support for Evren, the son of a mosque imam, and are bitter about blue-collar support for him. When the government began to crack down on labor, many thought that the trade unions got what they deserved.[32]

---

[28] *Turkey* (1983), p. 28.

[29] Mushtak Parker, "Turkey: Special Report," *Arabia: The Islamic World Review* 18 (February 1983): 37–42.

[30] *Turkish Daily News*, 22 April 1983; *Turkey* (1983) p. 28.

[31] *Wall Street Journal*, 4 November 1982; Sezer, "Turkey's Security Policies," p. 37; *Kuwait Times*, 7 March 1983; *Christian Science Monitor*, 5 April 1983; Kuniholm, *Origins of the Cold War in the Near East*, p. 146.

[32] Conversations with journalists and professors in Turkey, April 1983.

An explanation for their reaction may be found in one analyst's observation that, "whenever the Turkish peasant population has expressed its will in direct and open democracy, the result has astonished the elite controlling the monocentric political system."[33] Another explanation for the reaction may be that some of Turkey's elite misperceive the role that Islam will play in Turkey's political future.

Several scholars have argued that religion's role in Turkish politics should not be exaggerated nor its potential for alienating Turkey from the West overstated. It is not the *force* of religiosity—the intensification of belief—that has increased in Turkey, the Turkish scholar Serif Mardin argues, but the *scope* of religiosity—the range of social contexts within which religious considerations are relevant. The recent turmoil in Turkey, he believes, is not a sign of Islamic resurgence but rather a spin-off of the polarization between leftist and rightist ideologies, alternatives to religious protest that were made possible by the successful institutionalization of political democracy. The most important dimension of Islam's recent visibility in Turkey, Mardin adds, is that it meets an inner need for meaning not satisfied by the political culture. In this capacity, it serves as an important source of moral authority and cohesion for urban migrants and "transitionals." To the extent that President Evren encourages this role, while attempting to restore conditions for a viable democracy in Turkey, religion should be seen neither as a substitute for nor as a threat to the nationalist-oriented, secular political order but as a source of support for it. In this context, the anti-Western impulse in the revival of Islam, while it may reinforce Turkey's disenchantment with the West, does not necessarily dictate such a course.[34]

TURKEY AND NATO

Speculation about Turkey's political future raises questions about its military viability and its status within NATO. If Turkey and NATO both were important to each other in the 1950s, does that relationship still hold? Will it continue to be as important? Finally, what are the relationships between

---

[33]Mehmet, "Turkey in Crisis," pp. 59–60.

[34]Metin Heper, "Islam, Polity and Society in Turkey: A Middle Eastern Perspective," *Middle East Journal* 35 (Summer 1981): 345–63; Sezer, "Turkey's Security Policies," pp. 9, 42; Serif Mardin, "Religion and Politics in Modern Turkey," in *Islam in the Political Process*, ed. James Piscatori (Cambridge, MA: Cambridge University Press, 1983), pp. 138–59. Mango, in "Turkey: A Delicately Poised Ally," p. 63, has observed that, within the common framework of national self-assertion, different groupings in Turkey place different values on different foreign connections. The Center looks to the West, the Left to "socialist countries," and the extreme Right to the Muslim world. Mango notes that the key issue for Turkish opinion, however, is the willingness of foreign powers to accommodate Turkish national interests.

Turkey's European and Middle Eastern roles? Answers to these questions require an examination in some detail of the current military situation.

At present, in the event of an attack on NATO's central front, Turkey and Greece would tie down a considerable number of Warsaw Pact forces, some of which otherwise could be deployed on the central front. General Alexander Haig, in testimony before the U.S. Congress, asserted that in a war Turkey would tie down at least twenty Warsaw Pact divisions and possibly as many as thirty more along a Balkan front. Turkey's 569,000 troops, second in number only to those of the United States among NATO forces, together with Greek troops, comprise a total of twenty-five divisions which are designed primarily for defense. They face thirty-three relatively modernized Warsaw Pact divisions in the Balkans, while Turkey confronts another fifteen or so Soviet divisions in the Transcaucasia. The Warsaw Pact forces, designed for in-depth shock attacks, possess a mechanized and armored capability which outnumbers that of the Greeks and Turks by a 3-to-2 ratio.[35]

Warsaw Pact ground forces are supported by a Soviet Black Sea fleet of twenty-five submarines and forty-five to seventy-five surface ships (nearly one-third of the Soviet Union's major surface combatants) and 600 fairly advanced attack and interceptor aircraft. Deployment of the Fifth *Eskadra* in the Mediterranean since the mid-1960s gives the Soviet Union the capacity to balance the influence of the U.S. Sixth Fleet, whose ability to keep open the sea lines of communication is less certain, and to intimidate regional states. The Soviets also maintain approximately twenty Pacific fleet naval vessels in the Indian Ocean where they serve mainly as a monitoring function; only one-third of these are combatants.[36]

The Soviets see the Mediterranean as part of their own maritime defense perimeter; approximately 7,000 Soviet naval and merchants ships transited the Straits in 1978 (almost 20 per day), and the Soviets view the Straits as a key to their security. In the event of war, they would probably attempt to neutralize NATO's strategic position in Turkey, gain secure access to the Mediterranean and Indian Ocean (far more accessible after the reopening of the Suez Canal in June 1975), and interdict NATO's sea lines of communication in the Mediterranean and Persian Gulf, where Soviet air, surface, and

---

[35]Gürkan, "NATO, Turkey, and the Southern Flank," pp. 14, 55; *Turkey's Problems and Prospects*, p. 14; *Turkey, Greece, and NATO*, pp. 7, 15; discussions with officials in the departments of State and Defense.

[36]*Turkey's Problems and Prospects*, p. 15; *Turkey, Greece, and NATO*, p. 7; Sezer, "Turkey's Security Policies," p. 15; *Perspectives on NATO's Southern Flank*, pp. 2–3. Soviet "out-of-area" ship days, in which a navy is deployed beyond the naval operating and training areas of home waters, increased from 4,200 in 1965 to 16,500 in 1979 (approximately 45 ships in the area every day of the year); U.S. out-of-area ship days declined from 18,000 to 10,530 (approximately twenty-nine ships in the area every day of the year). Director of Naval Intelligence and chief of Information, Department of the Navy, *Soviet Naval Developments*, 2d ed. (Annapolis, MD: Nautical Aviation Publishing Company of America, 1981), pp. 16, 20. These statistics, however, account for neither the size nor the capability and function of different ships. A better index might be ten days or firepower capability per day.

submarine-launched cruise missiles pose a primary threat to all surface shipping. Because 90 percent of NATO's wartime supply to Greece and Turkey would be by sea, and because the Mediterranean acts as a lifeline to NATO's southern flank, the sea lines of communication in the eastern Mediterranean are vital. Control of the Mediterranean also protects NATO's southern flank from attack. As General Lyman L. Lemnitzer noted some years ago, the importance of the Mediterranean to Allied Command Europe's Southern Region cannot be overstated.[37] That remains certain today.

If the Warsaw Pact attacked Thrace (the southern flank's most vulnerable point) and the Straits and gained control of the Aegean, it would acquire a significant surge capability. This potential, in conjunction with Soviet capacity to project military power by air, underscores the importance of Turkey's control over the Straits and Turkish airspace and, hence, Turkey's role as a deterrent to Soviet actions in the Middle East as a whole.[38]

Major installations, whose value increased after the fall of the shah and the loss of monitoring stations in Iran, significantly enhance Turkey's value to the alliance. Incirlik provides forward basing for U.S. tactical fighter-bombers (F-4s deployed in rotation from Spain and Italy); Sinop, electromagnetic monitoring; Kargaburun, radio navigation; Belbasi, seismic data collection; Yumurtalik and Iskenderun, contingency storage of war reserve materials, ammunition, and fuel (20 percent of the U.S. Sixth Fleet's Mediterranean-based fuel is stored there); and Pirinclik, radar warning and space monitoring.

Along with numerous secondary facilities, which include fourteen NATO Air Defense Ground Environment early warning sites and twenty U.S. Defense Communication System terminal sites, these installations and facilities provide 25 percent of NATO's hard intelligence, some of which is not available from other sources on such matters as Soviet strategic nuclear activities, military systems development, and force readiness and movement. Facilities in Turkey serve a host of additional military functions as well. Tracking Soviet air attacks in the eastern Mediterranean, to cite but one example, would provide important protection for the Sixth Fleet, whose carrier battle forces are otherwise vulnerable to USSR-based bombers armed with cruise missiles and guided by satellite.[39]

---

[37]Gürkan, "NATO, Turkey, and the Southern Flank, pp. 38ff., esp. pp. 42, 47, 48; Rubinstein, *Soviet Policy*, pp. 25–26, 55, n51; Sezer, "Turkey's Security Policies," p. 53; Lyman L. Lemnitzer, "The Strategic Problems of NATO's Northern and Southern Flanks," *Orbis* 13, no. 1 (Spring 1969): 101, 105; Gürkan, in "NATO, Turkey, and the Southern Flank," p. 47, points out that "even the largest Soviet cruisers and the new Kuril [or *Kiev*] class ships are built with dimensions that enable them to pass through the Suez Canal."

[38]Lemnitzer, "Strategic Problems of NATO's Northern and Southern Flanks," p. 103; Gürkan, "NATO, Turkey and the Southern Flank," pp. 13, 48; Sezer, "Turkey's Security Policies," p. 21.

[39]*Turkey, Greece and NATO*, pp. 8–9; *Turkey's Problems and Prospects*, pp. 15–24; U.S., Congress, House, Committee on International Relations, *United States Military Installations and Objectives in the Mediterranean*, 95th Cong., 1st sess., 1977, pp. 37–47; U.S.,

Turkey's part in deterring Soviet adventurism in the eastern Mediterranean and Iran, and its potential role in protecting the West's access to oil in the Persian Gulf, were highlighted in October 1982 by a U.S.-Turkish colocator operating bases agreement.[40] If funding problems can be solved, this agreement will result in the modernization of ten Turkish airfields and the building of one or two new ones at Mus and Batman in eastern Turkey. With a length and width sufficient to accommodate long-range bombers and cargo planes, these airfields would enhance Turkey's deterrent role and perhaps help exercise a dissuasive influence on any aggressive Soviet intentions in the area. They would put modern NATO fighter-bombers within striking distance of the Transcaucasia and the Rapid Deployment Force (since 1 January 1983 the U.S. Central Command) within 700 miles of Abadan. That would be far closer to the head of the Persian Gulf than any of the facilities to which the Central Command now has access.

Even if these bases are designed primarily for NATO contingencies, and their use requires Turkish approval, their potential value in the event of a serious crisis means that whether or not they are actually used they will serve as an important deterrent, both to Soviet adventurism in Iran and to Soviet ambitions in the Persian Gulf.[41] Geoffrey Kemp, currently on the National Security Council staff, has argued, like many others, that NATO must redraw the maps of its strategic theaters to indicate the growing linkages between them, and that it must remove the artificial distinctions between NATO's borders and the Middle East for purposes of wartime contingencies.[42] That it will do so formally is unlikely. Germany and France consistently oppose such arrangements for constitutional and domestic-political reasons. Informally, however, and for planning purposes, NATO has begun to examine

---

Congress, Senate, Committee on Foreign Relations, *United States Foreign Policy Objectives and Overseas Military Installations*, 96th Cong., 1st sess., 1979, pp. 62–65; Rubinstein, *Soviet Policy*, p. 47; *Perspectives on NATO's Southern Flank,* p. 10; *Washington Post*, 17 May 1982. Because of the serious threat posed by Soviet maritime strike air forces based in the Crimea, U.S. naval plans, in the event of a serious contingency, call for the U.S. Sixth Fleet to withdraw from the eastern Mediterranean to the relative safety of Sicily and the Straits of Messina, where it could deal first with the subsurface threat and then work forward sequentially to deal with the surface and air threats in the eastern Mediterranean. Discussion with Vice Admiral William Henry Rowden, commander, U.S. Sixth Fleet, and NATO commander, Strike Force South, Bellagio, Italy, March 1983.

[40]Conversations with officials in the departments of State and Defense; *Washington Post*, 16 October and 1 November 1982. Under the colocator operating bases agreement, construction funds for the Mus and Batman bases are not programmed until fiscal 1984. Funding in that year is expected to be $69.2 million, of a programmed total of $135 million.

[41]For discussion of a somewhat analogous problem—Turkish willingness to support American U-2 flights—see the testimony of Warren Christopher, in U.S., Congress, House, Committee on Foreign Affairs, *Hearing on Supplemental Aid Requests for Fiscal Years 1979 and 1980 for Turkey and Oman,* 96th Cong., 1st sess., 1979. See also Harris, "View from Ankara," p. 134; and Gürkan, "NATO, Turkey, and the Southern Flank," p. 35.

[42]Geoffrey Kemp, "Defense Innovation and Geopolitics: From the Persian Gulf to Outer Space," in *National Security in the 1980's: From Weakness to Strength*, ed. W. Scott Thompson (San Francisco, CA: Institute for Contemporary Studies, 1980), p. 72.

"out-of-area" contingencies in order to overcome deficiencies in its capacity to react to emergencies, particularly in the Persian Gulf.[43]

If one had to rank the strategic importance of Turkey and Greece, geography and sheer numbers would give Turkey priority. This was true in the early 1950s, and it is even more true today. Arguments over the relative importance of either country, however, often ignore the great extent to which their strategic roles are complementary and interrelated: Turkey's role is to identify and counter a Soviet threat; in so doing, it helps to protect Greece. Greece supports operations in the Mediterranean which back up Turkey.[44] In short, neither would do very well without the other, and their joint efforts are crucial to the alliance.

Turkey's defense policy is predicated on deterrence, which makes NATO's security commitment and the threat of either horizontal or vertical escalation essential to Turkish security.[45] The Turks recognize, however, that their ability to resist intimidation must be grounded in their own internal resources, and that in the early stages of a war they would have to fight alone. They could not count on early reinforcements. The necessity of self-reliance, therefore, dictates a large defense establishment. The armed forces, based on conscription, not only provide assurance against the lurking uncertainty that NATO might fail to give Turkey adequate support in the event, for instance, of a limited conventional Soviet attack, but also serve an important socialization function.[46]

While the Turkish army is large, its equipment is antiquated, and its control systems are inefficient. As a modern military force, it is viewed by some military analysts as something less than reassuring. Its tanks, for example, are over twenty years old, most of its aircraft are also over twenty years old, and its naval vessels are more than thirty years old. This state of affairs is only partially the result of the embargo; in fact, the United States provided Turkey with considerable military assistance throughout the embargo years: $95 million in 1975 (the embargo was partially lifted on 6 October 1975 when arms worth $185 million and purchased before the embargo were released),

Turkish military equipment

---

[43]Conversations with officials in the departments of State and Defense. France has a permanent naval presence in the Indian Ocean of twelve to eighteen ships, periodically reinforced by one or two aircraft carriers; it also has 4,500 men stationed at Djibouti. In addition, British task forces of six ships deploy biannually in the Indian Ocean. U.S., Congress, House, Committee on Foreign Affairs, *NATO After Afghanistan*, 96th Cong., 2d sess., 27 October 1980, pp. 31ff. For U.S. desires to be free to conduct out-of-area operations in the Persian Gulf, see *New York Times*, 16 May 1982.

[44]*Turkey, Greece, and NATO*, pp. 6, 9, 14. The U.S. military installation in Suda Bay in Crete can accommodate almost the entire U.S. Sixth Fleet. *United States Military Installations and Objectives*, p. 32; *United States Foreign Policy Objectives and Installations*, p. 61; *Washington Post*, 15 May 1982.

[45]Taner Baytok, "Recent Developments in the Middle East and Southwest Asia: Impacts on Western Security," *NATO Review* 29, no. 4 (August 1981): 10–13.

[46]Gürkan, "NATO, Turkey, and the Southern Flank," pp. 17, 43–44, 48–49, 52. See also Sezer, "Turkey's Security Policies," pp. 26–27, n58.

$125 million in 1976, $125 million in 1977, and $175 million in 1978. More important to Turkey's deficiencies than a drop in military assistance was the fact that much next-generation equipment, which otherwise would have been available to the Turks, was devoured by the Vietnam War. Compounding this development was a rapid inflation in the cost of modernized weaponry and an American shift worldwide from grant to loan assistance.[47]

Turkey's modernization needs, emphasized first during the Carter administration, have been reemphasized by the Reagan administration. The Turks need everything but men to fulfill their NATO mission, including advanced fighter replacements for obsolete aircraft, an upgraded tank force, antitank and antiaircraft weaponry, and ASROC and *Harpoon* missiles.[48]

The Turkish air force's 325 combat aircraft include approximately 150 F-4s and F-5s and slightly over 120 serviceable F-100s and F-104s. The F-100s are, and the F-104s soon will be, obsolete. Fewer than one-half of Turkey's planes are combat ready, spare parts inventories are low or non-existent, and many planes have been cannibalized for parts; pilot training, consequently, has been limited. Turkey needs an advanced fighter aircraft and has sought a coproduction agreement for 290 F-16s. U.S. government officials have suggested a more reasonable number, given Turkey's needs, would be approximately 160. They believe that within a few years a ten-year program for coproduction/coassembly may be possible, although such a program would depend on arrangements to fund the approximately $3-billion cost.[49]

Turkey's 3,500 M-47 and M-48 A1 tank force is now antiquated. These tanks are fueled by gasoline and limited to a 160-mile range; they also have heavily worn 90mm. gun barrels that render them ineffective against modern tanks. The M-48s require, at a minimum, tank modernization kits that upgrade them to M-48 A5s, providing them with diesel engines, 105mm. guns, and improved fire control systems. Turkey's antitank weapons also are outdated; all but 1 percent consist of 3.5mm. rocket launchers and LAWs, with 500- and 250-meter ranges, respectively. The Turks require modern antitank weapons such as TOW missiles.[50]

The Turkish navy has fifteen destroyers, mostly *Gearing* and *Fletcher* classes, and all commissioned before 1946. It has no modern air defense; 89

---

[47]*Turkey, Greece, and NATO*, pp. 16–17; *Congressional-Executive Relations*, p. 25.

[48]*Turkey's Problems and Prospects*, p. 20.

[49]Ibid.; *Turkey, Greece, and NATO*, p. 19. The International Institute of Strategic Studies in the autumn of 1982 put the number of Turkey's combat aircraft at 402. *The Military Balance, 1982–1983* (London: IISS, 1982), p. 44. A recent report predicts that Turkey will soon announce the largest defense contract in its history. The contract will provide for 160 F-16s at $13–$15 million per aircraft and will be completed in the early 1990s. The package will probably include domestic production of subassembly parts, such as radar equipment, firing-control mechanisms, and range finders, and may include offset offers to sell Turkish export commodities through U.S. marketing channels. *Turkish Daily News*, 21 April 1983.

[50]*Turkey, Greece, and NATO*, p. 16. In June 1983 the Pentagon announced that it would sell 600 M-48 tank kits to Turkey for $155 million, *International Herald Tribune*, 9 June 1983.

percent of Turkey's short-range air defense is 1940 vintage or earlier, and most of it is unsupportable. All but three of its thirteen submarines, mostly *Guppy* class, are thirty-five years old.[51]

While the expense of upgrading Turkey's military forces only recently has been examined, preliminary estimates within the Pentagon of what it would cost merely to improve what Turkey had in 1980 approach $1 billion a year through 1985. In 1983 the Pentagon estimated that bringing Turkish forces up to minimum NATO standards would take $18 billion over thirteen years.[52] These estimates are subject to varying assumptions and are not free from manipulation.

TURKEY AND THE UNITED STATES

Given these assessments of Turkey's strategic importance and military deficiencies, and honoring the "best effort" commitment undertaken by the United States in the Defense and Economic Cooperation Agreement of 1980, the Reagan administration's fiscal year 1984 request for $930 million in military assistance to Turkey—$230 million in Military Assistance Program (MAP) grants, $525 million in Foreign Military Sales (FMS) credits, and $175 million in Economic Support Funds (ESF)—seems reasonable.[53] Whether it will appear so to Congress is another story. The de facto 7-to-10 ratio between aid to Greece and Turkey, while never agreed to by the executive and maintained with a certain slight of hand (ESF and supplemental requests for off-budget FMS credits are not counted in the ratio), has served in recent years as a reference point for congressional approval of military assistance to Greece and Turkey.[54]

The $930-million request for Turkey, once the $175 million in ESF is discounted, initially consisted of $755 million in military assistance. This figure was knocked down to $715 million in the House and Senate, and the administration was forced to come up with a budget amendment that raised

---

[51]Ibid.

[52]*Turkey, Greece, and NATO*, p. 18; *Turkey's Problems and Prospects*, p. 20; *Perspectives on NATO's Southern Flank*, p. 16.

[53]See U.S., Congress, House, Committee on Foreign Affairs, *Hearing on United States-Turkey Defense and Economic Cooperation Agreement*, 96th Cong., 2d sess., 1980, which includes a copy of the agreement. The DECA is a five-year executive agreement, renewable annually thereafter, that implements the NATO treaty. While the agreement provides for no specific amount, in it the United States undertakes to provide defense equipment, services, and training to the Turks; Turkey, in return, authorizes the United States to maintain forces and carry out military activities at specified installations in Turkey. The U.S.-Turkish Joint Commission was subsequently created as a mechanism for discussing how to use Turkey's available resources to meet its force goal objectives.

[54]Information obtained from the Office of Management and Budget and from the Congressional Research Service.

its request of $280 million in military assistance to Greece to $500 million, in effect restoring the 7-to-10 ratio.[55]

The administration still intends to hold additional military assistance to Greece as a quid pro quo for a base agreement with the Papandreou government.[56] Although congressional authorizations and domestic politics have made such tactics more difficult, they have by no means eliminated them. The money for Greece has yet to be appropriated. When Congress does so, the president still has considerable room for maneuver. If funds are not earmarked, the appropriation could be deferred or reprogrammed. If funds are earmarked, on the other hand, the president could seek to get the earmark waived. He also could defer the appropriation or even attempt to have it rescinded. In the interim, while the administration will have to lobby on behalf of the Turks, the Turks will have to await congressional appropriations, hoping that neither a delay in reaching a base agreement with Greece nor administration tactics in pursuit of such an agreement reduces the level of appropriations to Turkey. The administration also can attempt to make up any shortfalls through supplemental legislation.[57]

---

[55]Information obtained from the Office of Management and Budget, the Department of Defense, and the Congressional Research Service. See also *Congressional Quarterly Weekly Report* 41, no. 24 (18 June 1983): 1229–30. The fiscal 1983 request by the administration for aid to Turkey consisted of a military package of $465 million (a MAP grant of $300 million and an FMS credit of $165 million) and ESF of $350 million (a grant of $250 million and a loan of $100 million). In a continuing resolution, Congress appropriated $400 million in military assistance to Turkey. An additional supplemental request of $120 million (an FMS credit of $65 million and an ESF loan of $55 million), which would have brought the military package up to $465 million, was denied. As a result, the $400 million in military aid to Turkey, coupled with $280 million in military aid to Greece, maintained the 7-to-10 ratio.

These figures do not account for other sources of U.S. aid to Turkey, which are considerable. In fiscal 1983 they included $55 million in U.S. military construction funds, $34.6 million in navy ship transfers, $2.75 million in IMET; $2.3 million in DIC industrial machinery, and the U.S. share of a number of international programs and loans: World Bank project loans (20% of $700-$800 million), the UN Development Program (25% of $10 million), NATO Infrastructure (27.3% of $47 million), and NATO AWACS (27.3% of $1.3 million). In addition, Dependable Undertaking Funds totaling $244.77 million are available to the Turkish armed forces for allocation in the current fiscal year. These funds are drawn by the Defense Security Assistance Agency from future year MAP and FMS allocations through a cash flow financing procedure that is being used in 1983 to write FMS cases for the navy frigate program and the army tank modernization effort (see note 53 above). These figures put total U.S. aid to Turkey for fiscal 1983 at well over $1 billion, a total that is sometimes overlooked by elements within Turkey who are critical of U.S. policies toward their country.

[56]See *Christian Science Monitor*, 22 February 1983; and *Congressional Quarterly Weekly Report* 41, no. 13 (2 April 1983): 669.

[57]In the spring of 1983, Secretary of Defense Caspar Weinberger pushed hard for a supplemental request of $120 million to Turkey ($65 million in military assistance and $55 million in economic assistance), calling it "one of our most urgent priorities." *Christian Science Monitor*, 5 April 1983. It was rejected by a House Foreign Operations Subcommittee, and while its ultimate fate is unclear, it is not promising. *Congressional Quarterly Weekly Report* 41, no. 18 (7 May 1983): 887.

What all of these difficulties underscore is the extent to which congressional politics pose problems, not only for the administration's policies but also for the confidence of the Turkish government in the administration's capacity to sustain a program in Turkey that is not hostage to Turco-Greek differences.[58]

In spite of these problems, however, U.S. government officials believe that the special relationship between Turkey and the United States, while it is certain to lose some of its current euphoria, and while it will have its ups and downs, is on solid ground and will continue. One former U.S. ambassador to Turkey compared it to a marriage of over thirty years in which a couple has its differences but still remains committed, each to the other. From the Turkish side, Foreign Minister Ilter Turkmen in the spring of 1982 expressed the opinion that U.S.-Turkish ties had not been as good as they are now for years. The Turks appreciate American support during their domestic difficulties and contrast it with European criticism which caused the Turks to withdraw from the Council of Europe's Parliamentary Assembly in 1982.[59]

CONCLUSION: TURKEY AND THE FUTURE

In spite of Soviet moderation toward Turkey over the last seventeen years, traditional Turkish fears of the Soviet Union are periodically reinforced by Soviet military interventions in neighboring countries. As a result, alternatives to NATO—and the special relationship with the United States—are

---

[58]Turco-Greek problems are far-reaching and profound and periodically threaten to create serious problems for U.S.-Turkish relations [for example, see *Congressional Quarterly Weekly Report* 41, no. 19 (14 May 1983): 939; vol. 41, no. 24 (18 June 1983): 1229–30; and *International Herald Tribune*, 12, 13, 14–15, 19, 20 May 1983. They include differences over control of airspace in the Aegean, delimitation of the continental shelf, the limits of territorial waters, and last, but not least, Cyprus. While these issues are much too complex to examine in any detail here, failure to do so is not intended to minimize their importance; rather, it reflects the fact that some excellent recent monographs have examined Turco-Greek differences in depth and from several different viewpoints: Dennison J. Rusinow, "The Cyprus Deadlock: Forever or Another Day?" (American University Field Staff Reports, 1981/no. 11 Europe (Hanover, NH, 1981); Thanos Veremis, "Greek Security: Issues and Politics," *Adelphi Papers*, no. 179 (Winter 1982); and Andrew Wilson, "The Aegean Dispute," *Adelphi Papers*, no. 155 (Winter 1979–80). These monographs provide useful bibliographical references for those interested in further examining Turco-Greek differences.

[59]Interview, 25 February 1983 with James W. Spain, former U.S. ambassador to Turkey; *Christian Science Monitor*, 13 May 1983; Harris, "View from Ankara," p. 126. For a critical but nonetheless sympathetic U.S. view of Turkey's difficulties in effecting the transition back to democracy, see the text of Assistant Secretary of State for Human Rights and Humanitarian Affairs Elliot Abrams's testimony on 14 April 1983 before the Subcommittee on Human Rights and International Organizations of the House Foreign Affairs Committee. For a thoughtful assessment of the sensitive links between Turkey's political system and the U.S.-Turkish security relationship, see Ludington and Spain, "Dateline Turkey," p. 168.

not promising, at least as long as the Soviet Union remains a potential threat. The nonaligned movement is too fragmented; because of Cyprus, it is anti-Turkish. It is also unable to provide Turkey with the security or the technical and military assistance that Turkey requires.[60]

Turkey sees its future in conjunction with the West. If, as one analyst has suggested, a country's security policies are in part conditioned by the model of economic development that it chooses, Turkey's current economic policies and its choice of a mixed economy model clearly tie it to the West.[61] Differences with the West continue but should not be long lasting; European concerns over civil rights violations should abate as the generals carry out their pledges. On 7 November 1982 a constitution was approved. On 24 April 1983 a ban on political activity was lifted, a law governing the formation of political parties was instituted, and 16 May was set as the date for resumption of applications for new political parties. By July, four parties were in existence.[62] While a general election is scheduled for 6 November 1983, Evren's difficulties in reforming the structure of Turkish politics have led him to warn the country that the election may be delayed unless the politicians abide by his rules.[63] If all goes well by November or by the spring of 1984 at the latest, an election should be held and Turkey's relations with Europe should be on the mend.

It is true that the strategic environment in the region has been somewhat altered by a combination of increased Soviet strength and moderation, at least with respect to Turkey, as well as by a shift in the Middle East's economic influence in world affairs. If the Turks have accommodated themselves to these developments, however, there are limits to accommodation, not least of which is provided by a sense of history that breeds caution. Russia, whether

---

[60]Conversation with Turkish academics and officials in the military and foreign ministry. For discussion of these issues, see also Sezer, "Turkey's Security Policies," p. 38; Harris, "View from Ankara," p. 135; Gruen, "Ambivalence in the Alliance," p. 374; Rubinstein, *Soviet Policy*, pp. 40, 51, 146; Wilson, "The Aegean Dispute," p. 25; the opinions of Mehmet Ali Birand and Zeki Kuneralp, cited in Mango, "Understanding Turkey," p. 208; Vâli, *Bridge Across the Bosporus*, pp. 174–75; and *Turkey, Greece, and NATO*, pp. 28–30. Sezer observes that, while association with the United States entails risks, eliminating that association will not eliminate other risks and could increase them. The difficult choices confronting Turkey derive from the fact that like most countries it is neither totally secure nor totally threatened.

[61]Sezer, "Turkey's Security Policies," p. 39. For Turkey's recent plans to ban its old financial system of brokers and replace it with a regulated capital market, see *International Herald Tribune*, 16–17 April 1983.

[62]The political parties' law restricted participation—no old parties could be revived—and 480 former politicians were banned from political activity. By July 1983 five parties were actually founded, but one of them, the Grand Turk Party, was immediately dissolved, and sixteen former prominent politicians were taken into custody because it apparently sought to reincarnate the Justice Party, a development strongly opposed by Evren who wants to wipe the slate clean and reform the structures of Turkish politics. The media was not allowed to debate or comment upon the ban. *Christian Science Monitor*, 20 May 1983; *International Herald Tribune*, 1–3 June 1983.

[63]*International Herald Tribune*, 2 June 1983.

Christian or Communist, has demonstrated toward Turkey a relatively consistent attitude that is grounded in considerations of geopolitics.[64] Turkey's allies, too, have demonstrated a relatively consistent attitude based on the same considerations. Given the predispositions of the states that play a dominant role in the international balance of power, the NATO countries have needed Turkey and Turkey has needed them. Anti-imperialist rhetoric may confuse countries that do not border the Soviet Union; it does not confuse the Turks.

---

[64]Another source of Soviet concern over Turkey may be found in the 1979 Soviet census which revealed that 37.2 of 43.8 million Muslims in the USSR were of Turkic origin. Cited in Henze, *Goal: Destabilization*, p. 15. For more details of the problems posed for the Soviet Union by its national minorities, see Hélène Carrère d'Encausse, *Decline of an Empire: The Soviet Socialist Republics in Revolt*, trans. Martin Sokolinsky and Henry A. La Farge (New York: Newsweek Books, 1979).

Appendix: Comments by the Italian Chiefs of Staff

GENERAL UMBERTO CAPPUZZO: First, I must thank Professor Raimondo Luraghi, who inspired and organized this conference, for having invited me as guest speaker on this opening day. I would very much like to have been able to attend the entire conference which groups together internationally famous experts and analysts at the highest level. I am sure there were extremely interesting lessons to be learned, both for me personally because I am greatly interested in strategic problems and professionally as chief of staff of the Italian army.

Unfortunately, I can only make a brief appearance because the armed forces—and not just the Italian armed forces—are going through a delicate period, torn between insoluble budgetary problems and consequently equally arduous problems of efficiency and organization. It is a great pity as I am absolutely convinced not only of the validity of this conference, from which new ideas will emerge, but also of the importance of such meetings in providing a link between the cultural and operational fields. The operational aspect is often submerged by administrative problems and, therefore, there is a tendency to neglect the basis concepts. This then is another reason for thanking the people who envisaged and organized this conference. The fact that the chiefs of staff of the three services of the armed forces were invited separately to participate as guest speakers might lead one to think (but only if one were malicious) that together these three men would fight like Don Rodrigo's roosters, just to make a more or less appropriate historical reference to the region in which we are today.

Almost everyone knows, in fact, that strategic thinking, like all theoretical thinking, lends itself to adaptation. That is to say it is flexible and its final purpose is to demonstrate the absolute necessity of devoting major resources to one sector instead of another, or to the procurement of certain weapons systems rather than others. This, too, is very human and quite normal; it happens in every country, and the arguments among the chiefs of staff of the armed forces who want to obtain more money from the political authorities are a fact of life. This could probably be avoided by appointing someone as chief of defense who does not belong to any of the armed services, but this is an experiment that has never been carried out. Who knows what the results would be.

In any case, I want to assure you that I have no intention of trying to show you that NATO's southern flank is exclusively a land battleground, nor will I give examples from history to prove the theory that for thousands of years all invasions from the East have come via land, and that consequently we have to strengthen the army to the detriment of the navy and air force. Naturally, I hope my colleagues will be equally modest in their claims.

The very fact that this conference, focusing on the southern flank, is cosponsored by such a prestigious American institute as the Lemnitzer Center is of extreme importance and indicative of a different and more up-to-date approach to the Southern Region which, from a geopolitical and strategic point of view, is of critical importance. I am certainly not the only person to define it in these terms. In fact, we who belong to the southern flank believe that the Atlantic Alliance, from the very moment of its foundation, has owed us something that is not easily defined. Perhaps we could call it "due attention."

The political-strategic evaluations that have influenced not only the elaboration of NATO's military doctrine but also the political posture of the Atlantic Alliance originate, in my opinion, in a prevailingly continental approach that has automatically assigned to Europe's southern flank the exclusive role of right wing of the European defense deployment. At the same time, within the framework of the doctrine of containment, we find the basis for completing the external encirclement of the Communist world by political-military organizations comparable to NATO. And except for one key missing component that later proved to be vital for the stability and survival of these alliances, the others might have succeeded as well as NATO. Naturally, I am referring to the basic catalyst and unifying element in NATO: a common historical-cultural, social, and political heritage. Seen in perspective, and probably with hindsight, the encirclement operation (and notice I emphasize the word "encirclement" which is probably historically the term that first tsarist Russia and then the Soviet Union has feared most) has proved be an error of geopolitical judgment, as events have shown.

On the other hand, the Soviet Union, a continental nation par excellence, can operate through direct lines of action with a pendular movement in all directions. In this respect, the USSR has undertaken numerous ventures but has found Central Europe a compact and danger-filled obstacle. The Berlin Wall, erected after the events of 1961, seemed to be a symbolic materialization of the crystallization of this stalemate situation. If the events of the last thirty years are examined, it can be understood how the Soviets have been able to achieve objectives of exceptional importance without provoking a third world war and almost without becoming directly involved in armed conflicts. The last domino, to use a Cold War expression, Afghanistan was an exception to the rule and has shown that an unexpected capability for resistance exists with possibly sobering implications for future Soviet adventures.

I am not trying to be controversial but perhaps another approach by the West, taking into account global geopolitical reality without favoring the continental point of view, would have had different results. In other words, if the same attention, the same resources and energy had been devoted, politically and militarily, to southern Europe that were devoted to Central

Europe, the situation would probably not be what it is today, or at least the countries in this area would not have felt they were the "poor cousins" of the alliance.

Furthermore, I must add that the various assessments concerning southern Europe, repeated so often that they have become commonplace, are essentially correct. I would like to remind you of the most frequently used ones which concern the internal and international political instability of the subregions, the endemic economic weakness that creates a gap between their development and that of the more industrialized countries, the transient nature of the cohesive elements, and the difficulty in establishing a demarcation line, or front. But are these not all excellent reasons for a more active commitment and a more dynamic presence?

A quick review of the international events of the last twenty years will show that a more rapid evolution has occurred in our area of interest, that is to say on NATO's southern flank, than in Central Europe. Only certain events have to be listed: the increased Soviet naval presence in the Mediterranean, the oil crisis, the Pan-Islam revolution in Iran, the Iran-Iraqi war, Afghanistan, the situation in the Middle East and Lebanon as well as the disquietening conditions and attitudes in certain North African countries. These are all situations that demand a new understanding and evaluation vis-à-vis NATO's southern flank.

The overall situation in the Mediterranean must be reviewed, not as a reassessment of "direct" military threat which, in my opinion, has not been substantially modified by the presence of the Soviet fleet. Those ships, in any case, have neither the means nor any good reason to undertake military ventures of significant importance in a small and essentially hostile sea basin. It is more important to understand the fact that the Mediterranean area is directly connected to vital Western interests concerning both cooperation with the developing countries and as a basis for concrete acts of crisis management that lessen the danger of Western involvement in local conflicts, as well as for energy supplies and control of energy sources, which can no longer be considered a problem solely for NATO. An important first step forward was taken when Spain joined the NATO alliance, an event supported by the Italian government from the very start of the exploratory talks.

New ideas and new motivations are needed to rediscover the spirit of the Atlantic concept, the validity of which cannot and should not be questioned despite the "family rows." In particular, new momentum must be given to a global strategy in which the global aspect is not simply the simultaneous and coordinated use of all the available instruments (military, economic, social, ideological) but above all the unified and total involvement of every member of the alliance, with the conviction that it is in the area of NATO's southern flank that the future of the Western world, and possibly of humanity, will be played out.

ADMIRAL ANGELO MONASSI: It has been with a great deal of interest that I have followed the presentation of the conference. From my perspective, it has been instructive to see the continental situation presented from different angles and points of view, the result being a clear and well-defined picture of the primary factors affecting the Southern Region, especially the Mediterranean. I sense a mood inside the alliance, and in the Western world in general, that much has changed under the expansionist pressure of the Soviet Union from the situation we had for many years after the end of World War II. This mood exists at all levels and among different social and political environments an awareness exists and is growing with the slowness that is typical of democratic systems and vast multinational organizations.

Against the continental situation, precariously balanced on the bases of potential mutual destruction, it is evident that the situation, as developed in the Southern Region, is characterized by turbulence, complicated relationships, and unpredictable positions due to a general instability throughout the Mediterranean and adjacent areas. In addition, there remains the danger of flare-ups of local crises outside the geographic limits of the alliance, resulting in serious repercussions to the most vital interests of the West. What NATO has then, as already stressed by others at this conference, is a southern front and not a flank.

The front, with its expanded and direct contact with the economic arc, on the one hand, emphasizes the interdependence of external and internal factors and the strategic importance of peripheral areas. On the other hand, it demonstrates the Mediterranean's vital role as a sea line of communication, not only for NATO but also between the alliance and surrounding areas so important for the survival of the alliance.

This sea is long removed from NATO's mare nostrum of the 1950s. As recently as a few years ago, NATO still retained a certain supremacy over Warsaw Pact forces. This situation has now undergone a significant change, resulting in a definite alteration of the balance of power in the basin. It clearly demonstrates the Soviet Union's ever-increasing determination to assert itself forcefully as a naval power expanding the area of confrontation, thus enabling them to enlarge their strategic maritime area.

The Soviets have increased their presence not only in the Mediterranean but also in all other world waters through an expansion of their naval and merchant marine fleets. In the last few years the Soviet merchant fleet, with 23 million tons of ocean-going vessels operating profitably at competitive rates throughout the world, has reached a level that places it statistically among the largest in the world. It is an effective and highly visible instrument, carrying out an active commercial policy of new and multiple interests. The Soviet merchant marine is a necessary and well-adapted tool used to develop relationships of interdependence, sustain policies of penetration, and complete designs of political and military expansion under an ideological umbrella, or just in the name of the most pragmatic interest as the need arises.

In a surprisingly parallel development, the Soviet navy has grown to the point where it almost matches the largest of the Western navies. It displays a significant presence in the most critical operational areas, and it functions in such a multitude of roles that NATO's capability to defend the major sea lines of communication has become problematic, or at least extremely burdensome, particularly with regard to the most exposed areas, choke points, and high-density traffic areas like the Mediterranean.

This entirely new situation offers the Soviet Union more varied opportunities for exploitation. The danger to NATO, then, is clear: the possibility of an envelopment from the South, resulting in greater difficulty in obtaining raw materials and increased vulnerability of vital sea lines of communications which guarantee the flow of these supplies. Due to the confluence of such widespread vital interests and the possibility of a flare-up, owing to such changing and unpredictable events, this new situation deserves much more attention on the politico-military level by the alliance and by the individual nations affected as well as a different assessment based principally on current and more realistic thinking.

For Italy, which occupies the central position in this region and whose geography, historical importance, and current economic reality gives it the mission of functioning as the link between Central Europe and the southern belt countries, this responsibility is particularly evident. In the common interest of security, Italy is called to play an active role, assuming the reins of leadership that many Mediterranean countries acknowledge to be in accordance with Italy's political vocation and desire for peace. The necessity to expand its vision and political action, dictated by the current Mediterranean situation, with particular attention to the surrounding areas outside the Straits, has been recognized by Italy and is clearly demonstrated by its leaders who have authorized the assignment of Italian forces to such organizations as UNIFIL as well as to the Sinai and Lebanon Multinational Force. This conviction includes guaranteeing the neutrality of Malta.

This new line of thought requires a flexible military structure, readily available and characterized by high mobility that can concretely demonstrate involvement and political will to prevent and contain possible crises. This structure must be capable of adapting to the situation, ensuring maximum credibility, and adhering to political objectives.

In this regard, the Italian navy, while accomplishing its traditional missions, is aware of its new responsibilities stemming from the changed situation coupled with guidance coming from political authority. Toward this goal, it is continuing to put forth its best efforts and all available resources to accomplish the difficult task of improving the quality of equipment, facilities, and personnel training. The navy cannot ignore, however, needed improvements in its force structure. Any delay would be immediately obvious to those needing the services of the navy. In regard to a statement made by General Alberto Li Gobbi in his presentation, I would like to confirm his point concerning the defense budget cuts of the last two years brought about by the country's economic situation. These cuts have definitely caused a significant delay and postponement of programs for modernization and upgrading of equipment. In addition, the timetable has undergone yet another setback with regard to the achievement of the minimum level of operational tonnage and a balanced budget among the various components. Below this level the navy cannot guarantee credibility in its ability effectively to accomplish its mission in all situations.

If the budget situation has had a substantial negative impact on the completion schedule for its operational line and on the modernization of some serving equipment, this cannot be said for the training of navy personnel. Notwithstanding the technological complexity and sophistication of modern equipment installed on the navy's newer ships, in my estimation its personnel training continues to remain well above minimum acceptable levels.

GENERAL LAMBERTO BARTOLUCCI: Many thanks to the president and director of the Centro Studi sulla Difesa of the University of Genoa, who kindly gave me the opportunity to speak at such a distinguished forum. I have expressed already my satisfaction in seeing so many prominent scholars discuss, with impressive competence, specialized military problems.

In 1949 and during the early 1950s, the defense of Italy in relation to the defense of the NATO countries was significant only in its northern zone, that is, the Po Valley. In case of conflict, to lose the valley meant to withdraw NATO's southern based air assets back to the West. From the late 1950s Italy was forced to look at the situation in a different way, and the following points were open for consideration: the likely threat to NATO in the Mediterranean area; the role Italy could play to ensure peace and better security in the area; the kind of conflict that could be forced on us in this area in the immediate future; and the deployment that Italian air, land, and sea forces should attain in this context.

Anchored to Central Europe by the powerful chain of the Alps, Italy stretches far into the Mediterranean, dividing it into two basins and almost touching Africa. From a geographical point of view, Italy is part of Europe; from a political viewpoint, Italy *is* Europe. Africa is separated from Europe by the sea. If this is geographically true, in terms of geopolitics and geostrategy the situation changes completely: the Mediterranean becomes not an element of division but the main sea line of communication connecting Africa to the states of southern Europe.

The Mediterranean is a melting pot where major strategic lines of political, economic, and military power are found, lines that run from Washington to Moscow, from the industrialized North to the developing South, from the oil producers to oil consumers. This is an area where, besides permanent conflicts, such as the Middle East ones, terrorism, subversion, and guerrilla war, led by inside and remote agencies, have found fruitful soil in most countries. From the Strait of Gibraltar to the Persian Gulf, nearly sixty conflicts have been counted at various levels of severity since 1946 and about thirty coups, ten alone in Syria from 1949 to 1970. With all of this in mind how could it be possible to consider the Mediterranean area as just the southern flank and not a peculiarly delicate theater?

Air power in Italy is controlled by the Aeronautica Militare, the Italian Air Force, which can be regarded as one of the most significant expressions of the country and its people. It also can be regarded as a result of the nation's will to affirm its presence on the world scene as a sovereign state as well as a substantial component of the Atlantic Alliance and as a stabilizing factor of the global strategic balance.

The mission of the air force is a particularly difficult one considering the geographic dimensions and features of the theater in which it must operate. Among the many obstacles to its success are the limited financial resources that Italy can spare for expensive modernization programs. One can easily understand why the Italian Air Force is devoting so much attention to this area. It is one of political instability, where in the last few years Soviet strategy had been changing through an increased military presence and infiltration into several North African and Middle Eastern countries. We may well observe that both Soviet military forces and new military bases are greater than those needed for any reasonable defense. Particularly, at any moment they can receive far larger forces without any logistical limit.

Given a threat coming potentially from any direction, the air force has to deploy forces that can resist attack in Italy's northeastern and southern theaters with equal ability. The costs in terms of men and machines are high. Since Italy's northeastern border remains the only way through which it might be invaded by land forces, the Italian Air Force must keep its deployment in the northeast unaffected by the threat that is growing in the South, where air attacks and naval actions may be expected against the maritime traffic of our country and of our allies.

In the South the assets of the air force must be devoted not only to an air defense sector covering a huge area of responsibility but also must make a substantial contribution to the Italian navy. This means that air power must deal with the enemy's sea as well as air forces, the latter both in their own bases or when airborne. Should the enemy have its own bases on the North African coasts, the danger for Italy's military and merchant fleets will be substantially increased. It is no secret that in the North African area well-protected air force bases are being continuously built, capable of hosting squadrons of military aircraft of the most recent generations, well able to reach Italian targets even in a low-altitude profile. Consequently, Italy has positioned its air forces near the zone where its naval units will operate. Also being studied and implemented are possible actions that might enable Italy's air force to control the southern sector in the most effective way and to react to any aggression from the very beginning.

As early as the 1960s plans had been made to adjust the Italian Air Force gradually and systematically to new situations to cope with mounting tensions in the Mediterranean and to contribute to the building of peace and security in the area. Italy's aim is to deter potential threatening nations from starting dangerous adventures. In this endeavor, it is currently implementing plans to shift several squadrons south, with others soon to follow; increase the flexibility of forces in order to act quickly if and when it should be necessary; enlarge radar networks in the Mediterranean sector, which will soon have AWACS planes and mobile radar stations there in case of a sudden breakdown in some part of its system; continue to increase potential in the South without sacrificing the northeast posture; and finally also to promote further land-based support and improve the capability of new weapons systems as well as to use in-flight refueling and to acquire adequate standoff weaponry.

For the Italian Air Force all this presents a major effort on multiple fronts: supplies, weapons programs, and supports. It not only requires new funds but also adequate training and efficiency. Toward this goal something new is needed, perhaps a more substantial dimension of Italy's forces, which in turn raises certain problems with the defense budget. However, Italy is proceeding with clear and straightforward ideas to maximize the money spent by means of stable and coordinated programs, keeping in mind the need to integrate the several branches and to coordinate its weapons systems and tactics. Only through united efforts will Italy be ready to respond adequately to the new challenges. On this assumption the air force proceeds to do its duty.

Index